The Planet Pluto

PERGAMON TITLES OF RELATED INTEREST

BOOKS

BATTEN   Binary and Multiple Systems of Stars

CLARK & STEPHENSON   The Historical Supernovae

GLASBY   Nebular Variables

GURZADYAN   Flare Stars

HEY   The Radio Universe, 2nd Edition

HILLAS   Cosmic Rays

MEADOWS   Stellar Evolution, 2nd Edition

PACHOLCZYK   Radio Galaxies

PAPADOPOULOS   True Visual Magnitude Photographic Star Atlas (3 volumes)

REDDISH   Stellar Formation

SAHADE & WOOD   Interacting Binary Stars

SOLOMON & EDMUNDS   Giant Molecular Clouds in the Galaxy

JOURNALS

CHINESE ASTRONOMY
PLANETARY AND SPACE SCIENCE
VISTAS IN ASTRONOMY

THE PLANET PLUTO

A.J. Whyte
Director,
Ellerslie Observatory

Pergamon Press
Toronto □ Oxford □ New York □ Sydney □ Frankfurt □ Paris

CANADA	Pergamon of Canada, Suite 104, 150 Consumers Road, Willowdale, Ontario M2J 1P9, Canada
U.K.	Pergamon Press Ltd., Headington Hill Hall, Oxford OX3 0BW, England
U.S.A.	Pergamon Press Inc., Maxwell House, Fairview Park, Elmsford, New York, 10523, U.S.A.
AUSTRALIA	Pergamon Press (Aust.) Pty. Ltd., P.O. Box 544, Potts Point, N.S.W. 2011, Australia
FRANCE	Pergamon Press SARL, 24 rue des Ecoles, 75240 Paris, Cedex 05, France
FEDERAL REPUBLIC OF GERMANY	Pergamon Press GmbH, 6242 Kronberg-Taunus, Pferdstrasse 1, Federal Republic of Germany

Copyright © 1980 Pergamon of Canada Ltd.

Canadian Cataloguing in Publication Data

Whyte, Anthony J., 1951-
　The planet Pluto

　Bibliography: p.
　Includes index.
　ISBN 0-08-024648-6

1. Pluto (Planet). I. Title.

QB701.W48   523.4'82   C79-094935-0

Cover photograph courtesy of:
U.S. Naval Observatory, Flagstaff Station;
enlargement by Stew Jones,
Lowell Observatory, Flagstaff, Arizona.

All Rights Reserved. No part of this publication may be reproduced, stored in a retrieval system or transmitted in any form or by any means: electronic, electrostatic, magnetic tape, mechanical, photocopying, recording or otherwise, without permission in writing from the copyright holders.

In order to make this volume available as economically and as rapidly as possible the authors' typescripts have been reproduced in their original forms. This method unfortunately has its typographical limitations but it is hoped that they in no way distract the reader.

Printed in Canada

To my friend and mentor,

HARRY A. WISE

Contents

	Preface	ix
CHAPTER 1.	The Expansion of the Solar System, 1781 - 1846.	1
CHAPTER 2.	The Search for Planet 'X', 1847 - 1927.	16
CHAPTER 3.	The Discovery of Pluto, 1928 - 1930.	32
CHAPTER 4.	Filling in the Blanks, I, 1931 - 1956.	44
CHAPTER 5.	Filling in the Blanks, II, 1957 - 1972.	64
CHAPTER 6.	Cold Moonlight, 1973 - 1979.	95
CHAPTER 7.	Full Circle, 1980 - 2178.	127
	Bibliography	130
	Name Index	142
	Subject Index	145

Preface

This book has been written in the hope that it will fill a long-standing gap on the shelves of the astronomer's library. *The Planet Pluto* joins a small family of books with similar titles and purposes, including: W. Sandner's *The Planet Mercury*, B.M. Peek's *The Planet Jupiter*, and A. Alexander's *The Planet Saturn*. Hitherto no attempt has been made to present in a single volume an account of the search for a planet exterior to Neptune, the discovery of Pluto, and what we know about this remote world; until very recently the last mentioned was very little indeed. However, with the recent discovery of a satellite to Pluto on the eve of the 50th anniversary of the planet's discovery, our knowledge of Pluto has increased significantly; sufficiently so, that a book such as *The Planet Pluto* is not only feasible, but desirable.

It is hoped that *The Planet Pluto* provides a readable account of the search for the unknown planet exerting its influence on the motions of the other planets of the solar system and that it conveys some of the tremendous public excitement and scientific interest aroused by the discovery of Pluto by Clyde W. Tombaugh in 1930. As for presenting a useful précis of the scientific literature on Pluto, it is hoped that the book's extensive bibliography will prove especially valuable to researchers of the planet. Should any significant contributions to our knowledge of Pluto have been overlooked, this has occurred unintentionally.

Anthony J. Whyte,
July, 1979.

Acknowledgements

I am grateful for the support and assistance provided by a large number of individuals and institutions. Special appreciation is owed to the following:

Merle Martin; James W. Christy; Ian Halliday; John Marelli; Brian Marsden; Claus Oesterwinter; Marguerite Whyte; James G. Williams; The University of Alberta, Physical Sciences Library; The United States Naval Observatory Library; The Ellerslie Observatory, Board of Directors.

Chapter 1
The Expansion of the Solar System
1781-1846

Since antiquity there had been only five known planets besides the earth and it had been on this diminutive family of planets that successive generations of astronomers had based their cosmologies, progressing over the centuries from Ptolemaic geocentric epicycles to heliocentric Keplerian ellipses governed by Newtonian laws of gravity. By the middle of the eighteenth century astronomers had described the known solar system with great fidelity and accuracy and there apparently remained little to learn. Therefore the discovery of a seventh planet by William Herschel and the consequent doubling of our solar system's breadth came as a complete surprise.

In 1781 Herschel had initiated what was to be his life's work: a complete and systematic survey of the heavens involving the telescopic observation and description of the position and appearance of every visible object. On the night of March 13 Herschel, using his favourite 16-cm reflector, noticed a curious non-stellar object near the star H Geminorum. He suspected it to be a comet; in his own words:

> The power I had on when I first saw the comet was 227. From experience I knew that the diameters of the fixed stars are not proportionately magnified with higher powers, as the planets are; therefore, I now put on the powers of 460 and 932, and found the diameter of the comet increased in proportion to the power, as it ought to be, on a supposition of its not being a fixed star, while the diameters of the stars to which I compared it were not increased in the same ratio. Moreover, the comet being magnified much beyond what its light would admit of, appeared hazy and ill-defined with these great powers, while the stars preserved that luster and distinctness which from many thousand observations I know they would retain.

On March 19, Herschel found that the object was moving eastward among the stars confirming his suspicions that it was a comet although he was later to note, on April 6, that the object appeared perfectly sharp upon the edges, without any appearance of a tail. The possibility that the object was a planet, though perfectly compatible with the visual evidence, did not at first occur to Herschel. Indeed, only Nevil Maskelyne, the Astronomer Royal, among the three men Herschel first informed of his discovery, suspected that the object might be a planet. On April 23, Maskelyne suggested the planetary possibility in a letter to Herschel. Nevertheless, Herschel (1781) read a paper entitled simply "Account of a Comet" before the Royal Society on April 26, stating incorrectly that the object's daily parallax was between 10" and 20" placing its position within the Earth's orbit.

Upon being notified by Maskelyne of the discovery many continental astronomers including Charles Messier, Joseph Jerome Le Francais de Lalande, Pierre Mechain, Johann Bode and others undertook observation of the object in an attempt to determine its motion. As soon as enough observations were accumulated several unsuccessful attempts were made to derive a parabolic (cometary) orbit matching the object's actual motion. The incompatibility of this motion with cometary-type orbits was finally forced upon astronomers after repeated failures at reconciliation. Messier, writing to Herschel, commented that the latter's "comet" bore no resemblance at all to any comets he had discovered which at that time numbered eighteen.

Abandonment of the comet hypothesis followed quickly after Anders Johann Lexell (1781) summed up the non-cometary characteristics of the object and computed a circular (planetary) orbit for it. Differing sets of circular elements satisfying the short observed path thus far travelled by the object were soon published in Russia, France, and Germany. It was obvious that observations covering a greater span of time were needed before a more accurate orbit could be established.
Wide acceptance of the planetary nature and hence, the significance, of Herschel's discovery brought with it acclaim. In 1782 Herschel was appointed court astronomer to George III of England. In gratitude Herschel proposed the name "Georgium Sidus" for the new planet, while Lalande nominated "Herschel" and Bode suggested "Uranus" in keeping with the mythological tradition of the other planets' names. For some sixty years after it's discovery the new planet went by three names until "Uranus" prevailed.

The Expansion of the Solar System

By 1783 the preliminary circular elements for Uranus had been superceded by the collaborative calculation, by Laplace and Mechain, of the first set of elliptical elements. Tables of the new planet appeared in the German and French ephemerides for 1781 which were published in 1784. Recognizing the limitations placed on the accuracy of such tables by the small number of observations of Uranus since its discovery, Bode sought to extend the span of observations by searching prediscovery catalogues in which Uranus might have been recorded as a star. In this Bode was most successful and in August, 1781 he found that Tobias Mayer had mistakenly recorded Uranus as a star in Aquarius on September 25, 1756. Three years later, aided by Placidus Fixlmillner, Bode found that the first Astronomer Royal, John Flamsteed, had observed Uranus on December 23, 1690 in Taurus.

The finding of Flamsteed's 1690 observation meant that positions were now available for Uranus for more than one revolution. Fixlmillner (1787), using the observations of 1690 and 1756 plus those of 1781 and 1783, was able to calculate the following set of elliptical elements (Table 1):

TABLE 1. Fixlmillner's Elliptical Elements for Uranus

Epoch, January 1, 1784

Mean anomaly	297°	9'	25"
Longitude of perihelion	167	31	33
Longitude of ascending node	72	50	50
Inclination	0	46	20
Tropical period (days)		30587.37	
Mean distance (A.U.)		19.18254	
Eccentricity		0.0461183	
Mean daily tropical motion		42".3704	

By 1788 however, it was apparent that something was amiss for, in that year, the tables were in significant discordance with the observed positions of Uranus. Fixlmillner, in reworking his elements and tables, found that he could not reconcile Flamsteed's 1690 observation with contemporary observations so he omitted it in his calculations. Other theoreticians including Barnaba Oriani, Joseph Lalande and Jean Delambre attempted to overcome the discrepancies by allowing for the perturbation effects of Saturn and Jupiter upon Uranus. Delambre's tables, in particular, succeeded in reconciling all contemporary observations with those of Mayer and Flamsteed.

For the next thirty years interest in Uranus waned as astronomers, lulled by Delambre's apparently correct tables, turned their attention to other matters of interest. The number of recorded observations of Uranus declined precipitously while, conversely, more and more pre-discovery observations of the planet were uncovered. In 1788 Pierre Lemonnier found that he had mistaken Uranus for a star in 1764 and 1769. Friedrich Bessel (1818) found that James Bradley had recorded Uranus as a star on December 3, 1753. Johann Burckhardt discovered that Flamsteed had observed Uranus not only in 1690, but also on April 2, 1712 and April 29, 1715. And in 1820 Alexis Bouvard, in going through the notes of the now deceased Lemonnier, found ten more old observations of Uranus. In January, 1769 alone Lemonnier had observed Uranus six times in nine days without suspecting its non-stellar nature. Table 2 lists the prediscovery observations of Uranus as of 1820.

TABLE 2. Pre-discovery Observations of Uranus

Date	Observer
Dec. 23, 1690	Flamsteed
April 2, 1712	"
April 29, 1715	"
Oct. 14, 1750	Lemonnier
Dec. 3, 1750	"
Dec. 3, 1753	Bradley
Sept. 25, 1756	Mayer
Jan. 15, 1764	Lemonnier
Dec. 27, 1768	"
Dec. 30, 1768	"
Jan. 15, 1769	"
Jan. 16, 1769	"
Jan. 20, 1769	"
Jan. 21, 1769	"
Jan. 22, 1769	"
Jan. 23, 1769	"
Dec. 18, 1771	"

By now even Delambre's tables of 1790, like those earlier of Fixlmillner, had begun to seriously disagree with observations of Uranus's position. Accordingly, Alexis Bouvard (1821) undertook to correct Delambre's errors and calculate new tables for all three outer planets. He successfully completed the corrections for Jupiter and Saturn, but encountered the same difficulty with Uranus that had earlier troubled Fixlmillner; namely, how to reconcile the pre-discovery observations with the modern ones. But whereas Fixlmillner had only two old observations to cope with, one of which he deleted entirely, Bouvard had the seventeen listed above (Table 2). Bouvard found the problem insolvable as he admitted in the introduction to his new tables:

> The construction of the tables of Uranus involves this alternative:- if we combine the ancient observations with the modern ones, the first will be adequately represented, but the second will not be described within their known precise tolerances;- while if we reject the ancient positions and retain only modern observations, the resulting tables will accurately represent the latter, but will not satisfy the old figures. We must choose between the two courses. I have adopted the second as combining the most probabilities in favor of truth, and I leave to the future the task of discovering whether the difficulty of reconciling the two systems results from the inaccuracy of the ancient observations, or whether it depends on some extraneous and unknown influence which may have acted on the planet[1].

Bouvard's rejection of the pre-discovery observations implied that the measurements of the old observers, including two Astronomers Royal, were so inaccurate as to be utterly unreliable. Such condemnation did not pass completely unchallenged by Bouvard's contemporaries. Urbain Jean Joseph Leverrier later (1846a) recorded his impression at the time of the validity of Bouvard's claims:

> One should note that, though the observations of Flamsteed, Bradley, Mayer, and Lemmonier are not as exact as those of the astronomers of our epoch, one may not with any plausibility be allowed to consider them infested with such enormous errors as those of which the present tables accuse them. The author of these tables (Bouvard) actually suggests, however, that this is his opinion[2].

---

[1] Bouvard, A., *Tables astronomique* .... Paris, 1821.
[2] Leverrier, U.J.J., Recherches sur les mouvements d'Uranus. *C.R. Acad. Sci. (Paris)*, 22, 907 (1846)

Bessel detected a number of theoretical errors in Bouvard's calculations which, though minor, did nothing to allay his skepticism. Bessel also began new observations of Uranus in 1823, checking for any discrepancies between these and Bouvard's predicted positions. By the middle of 1825 the observed positions of Uranus had begun to diverge significantly, though not alarmingly, from Bouvard's tables. The disagreement grew over the next year. Of fifty four observations recorded by Bonifacius Schwarzenbrunner at the Kremsmunster Observatory around the opposition of Uranus during July, 1826, less than half were within an acceptable 5" variance from the tables.

Then inexplicably, the observed heliocentric longitude of Uranus which had formerly remained steadily in advance of the calculated longitude, began to decrease until, in 1829-30, the tabular and observed longitudes matched. But instead of keeping pace with its calculated orbit Uranus began to lag further and further behind until the tables were in total disarray. In a report to the British Association for the Advancement of Science, George Biddell Airy (1832), soon to be Astronomer Royal, noted that Bouvard's eleven-year old tables were in error in geocentric longitude by almost a half-minute of arc.

Following his report to the B.A.A.A., Airy received a letter from a British amateur astronomer, the Reverend Dr. Thomas Hussey, who sought Airy's opinion on the likelihood of the irregularities in Uranus's motion being caused by an unknown planet and whether it would be worth his while to search for it. Airy's negative reply discouraged Hussey from undertaking his search, but the possibility of the existence of an eighth planet was raised several times over the next few years by professional astronomers. Elix Valz (1835) and Friedrich Nicolai (1836), directors of the Mannheim Observatory, both suggested the presence of another planet beyond Uranus as a result of their investigation of perturbations in the path of Halley's comet.

The director of the Palermo Observatory, Niccolo Cacciatore (1836), reported observing a moving star in May, 1835 on three occasions before it disappeared in the twilight. He suspected that it was located beyond Uranus because of its slow motion.

Early in 1841 a brilliant young student at Cambridge University, John Couch Adams, happened upon Airy's *Report* of 1831-1832 to the B.A.A.A. in a Cambridge bookshop. Immediately his abiding interest in astronomy was intrigued by Airy's account of the discrepancies in the motion of Uranus and he wrote a memorandum:

> Formed a design, in the beginning of this week, of investigating, as soon as possible after taking my degree, the irregularities in the motion of Uranus, wh. are yet unaccounted for; in order to find whether they may be attributed to the action of an undiscovered planet beyond it; and if possible thence to determine the elements of its orbit, &c approximately, wh. wd. probably lead to its discovery.

The demands of the Cambridge curriculum prevented Adams from tackling the Uranus problem until after the completion of his examinations and subsequent graduation with highest honours. Then, during a much needed vacation, amid the relaxed surroundings of his family's home in Cornwall, Adams began his long-contemplated investigation of Uranus. By October, 1843 his computations, though preliminary and approximate, supported his belief in a new planet as the source of the perturbations in Uranus's orbit.

Adams's first analysis did not include any pre-discovery observations of Uranus. For his calculations he had used Bouvard's tables for the years 1781 to 1821 with later data from Airy's observations published in *Astronomische Nachrichten* and from the records of the Cambridge and Greenwich Observatories. Adams started his second, revised analysis around New Year's 1844. He first asked Professor James Challis to provide him with unpublished Greenwich planetary observations for the period 1818-1826 which, in his first analysis, had been the years of greatest discrepancies in Uranus's position. Challis obligingly obtained the necessary data from Airy. Adams based his revised equations on a complete series of ancient and modern observations. He retained his initial mean orbital radius value of 38.4 astronomical units, but gave up the simplicity of a circular orbit by incorporating terms to take into account the probable eccentricity of the new planet's orbit. However various interruptions prevented Adams from completing his work until September, 1845.

When Adams showed his results to Challis the latter urged him to send them to Airy, the Astronomer Royal. Adams decided to deliver them personally on his way home to Cornwall and Challis provided him with a letter of introduction to Airy. However Airy was in Paris at the time attending a meeting so Adams called again on his return from Cornwall. Unfortunately through a series of misunderstandings Adams never got to see Airy and returned to Cambridge feeling piqued, but leaving a summary of his solution to the problem of Uranus:

> According to my calculations, the observed irregularities in the motion of *Uranus* may be accounted for by supposing the existance of an exterior planet, the mass and orbit of which are as follows:-

Mean Distance (assumed nearly in accordance with Bode's law)	38.4
Mean Sidereal Motion in 365.25 days	1°30'.9
Mean Longitude, 1st October, 1845	323°34'
Longitude of Perihelion	315°55'
Eccentricity	0.1610
Mass (that of the Sun being unity)	0.0001656

For the modern observations I have used the method of normal places, taking the mean of the tabular errors, as given by observations near three consecutive oppositions, to correspond with the mean of the times; and the Greenwich observations have been used down to 1830: since which, the Cambridge and Greenwich observations, and those given in the *Astronomische Nachrichten* have been made use of. The following are the remaining errors of mean longitude:-

Observation - Theory

1780	+0".27	1801	-0".04	1822	+0".30
1783	-0.23	1804	-0.04	1825	+1.92
1786	-0.96	1807	-0.21	1828	+2.25
1789	+1.82	1810	+0.56	1831	-1.06
1792	-0.91	1813	-0.94	1834	-1.44
1795	+0.09	1816	-0.31	1837	-1.62
1798	-0.99	1819	-2.00	1840	+1.73

The error for 1780 is concluded from that for 1781 given by observation, compared with those of four or five following years, and also with Lemonnier's observations in 1769 and 1771. For the ancient observations, the following are the remaining errors:-

Observation - Theory

1690	44".4	1750	-1".6	1763	-5".1
1712	+6.7	1753	+5.7	1769	+0.6
1715	-6.8	1755	-4.0	1771	+11.8

> The errors are small, except for Flamsteed's observation of 1690. This being an isolated observation, very distant from the rest, I thought it best not to use it in forming the equations of condition. It is not improbable, however, that this error might be destroyed by a small change in the assumed mean motion of the planet[3].

Unaccountably Airy reacted negatively to Adams's paper - an attitude that was later to have major consequences. The response of the Reverend William Dawes to Adams's results upon being shown them by Airy was in marked contrast to the latter's. Dawes wrote at once to his friend, William Lassell, sending him Adams's positions and urging him to look for the new planet with his 60 cm. reflector. Unfortunately Lassell mislaid Dawes's note before being able to act upon it.

When Airy finally replied to Adams's paper on November 5, 1845 he did so by couching his rejection in a question. Airy asked whether the "assumed" perturbation of Uranus would explain the error of the radius vector of Uranus which was by then very large. Adams felt put off by Airy's question and was annoyed by his repeated use of the word "assumed" and did not reply for over a year. In his letter Adams explained his reasoning:

> ...For several years past the observed place of Uranus has been falling more and more behind its tabular place. In other words the real angular motion of Uranus is considerably slower than that given by the tables. This appeared to me to show clearly that the tabular radius vector would be considerably increased by any theory which represents the motion in longitude, for the variation in the second member of the equation

$$r^2 \cdot d\theta/dt = \sqrt{a(1 - e^2)} \quad (1)$$

---

[3]Airy, G.B., Account of some circumstances.... *Mon. Not. R. astr. Soc.*, 7, 123 (1846). By permission of the Royal Astronomical Society.

is very small. Accordingly I found that if I simply
corrected the elliptic elements, so as to satisfy the
modern observations as nearly as possible, without taking
into account any additional perturbations, the corresponding increase in the radius vector would not be very
different from that given by my actual theory[4].

Challis (1846) dismissed Airy's objection about the radius
vector in a letter to the editor of the *Athenaeum* for December
17, 1846. Airy promptly defended his position in a letter to
Challis, stating that there was no *a priori* reason for believ-
ing that a hypothesis which would explain the error of longitude will also explain the error of radius vector. Which is
true since, as Littlewood (1957) has pointed out, in Adams's
equation (1), the variation in the angular momentum
($r^2 \cdot d\theta/dt$) is proportionately of the same order as the error
in angular position (θ = longitude); correction of the longitude did not automatically correct the radius vector.

Whatever the validity of Airy's objections, the net result of
his obfuscation was to fritter away Adams's - and English
astronomy's - lead over Urbain Leverrier. For on November 10,
1845 Leverrier (1845) presented to the French Academy of
Sciences the results of his rigorous analysis of the motion of
Uranus. Leverrier's results, which could be applied to any
epoch, were based upon observations made from 1781 to 1801 at
Greenwich Observatory and from 1801 to 1845 at the Observa-
toire Royal in Paris. In his paper Leverrier demonstrated
that tables calculated from hypothetical elements for Uranus
based on observations for the period 1790 to 1820 would show
an error of more than 40" of arc at the opposition of 1845.
He concluded that: "Such is the order of the error in the
actual tables, except that the discrepancy is larger, and the
surplus can be attributed to outside causes, whose effect I
will evaluate in a second Memoir." Leverrier's paper reached
England in December and received a much more favourable re-
ception from Airy than had Adams's.

Leverrier worked on his promised second memoir (1846a) during
the first half of 1846 and presented it to the Academy of
Sciences on June 1. In his paper Leverrier concluded that the

---

[4]Smart, W.M., John Couch Adams and the discovery of Neptune.
Occasional Notes of the R. astr. Soc., 2, 56 (1947). By
permission of the Royal Astronomical Society.

observed irregularities in Uranus's orbit could best be attributed to the perturbing effect of an unknown planet. Next he showed how if this hypothetical planet was to have only an effect upon Uranus and not Saturn as well, it would have to be situated beyond Uranus. Then he assumed a mean orbital radius for the new planet about twice that of Uranus, basing this guess upon Bode's law (Table 3).

Noting that the orbits of Jupiter, Saturn and Uranus had only slight inclinations to the ecliptic, Leverrier proposed the same condition for the orbit of the hypothetical planet. He found his final necessary coordinate, the longitude at any epoch, by mathematically comparing the effect upon Uranus of various combinations of mean longitude and assumed mass for the hypothetical planet. After an exhaustive analysis of the numerous possibilities he eventually concluded: "there was only one region of the ecliptic where the perturbing planet could be located in order to account for the irregularities of Uranus; the mean longitude of the planet on the 1st of January, 1800 must have been between 243° and 252°." After further refinement and correction for epoch, Leverrier arrived at a heliocentric longitude of 325°, for epoch January 1, 1847.

Airy, who received Leverrier's second memoir on Uranus in June, 1846 and spoke enthusiastically of "the very close coincidence between the results of Mr. Adams's and M. Leverrier's investigations of the place of the supposed planet disturbing Uranus," at a special meeting of the Board of Visitors to Greenwich Observatory. Although Airy now believed in the possibility of a planet exterior to Uranus he was no more anxious to undertake a search for it than he had been before despite Leverrier's request that he do so. Yet Airy abruptly had a change of heart during the first week of July, 1846. Possibly realizing the storm that would erupt if it was found out that despite having been in possession of Adams's predictions for over eight months he had, through his inaction, allowed someone else to discover the new planet first. Hence in a letter dated July 9, followed by an even more urgent one on July 13, Airy asked Challis to begin searching for the planet with the Cambridge Observatory's 30 cm. refractor.

While Challis was searching for the new planet Leverrier (1846b) presented his third memoir entitled, "Sur la planete qui produit les anomalies observees dans le mouvement d'Uranus. - Determination de sa masse, de son orbite et de sa position actuelle," to the Academy of Sciences on August 31.

TABLE 3. Bode's Law of Planetary Radii (in A.U.)

	Mercury	Venus	Earth	Mars	Ceres	Jupiter	Saturn	Uranus	?
	.4	.4	.4	.4	.4	.4	.4	.4	.4
	+0.0	.3	.6	1.2	2.4	4.8	9.6	19.2	38.4
Law:	.4	.7	1.0	1.6	2.8	5.2	10.0	19.6	38.8
Observation:	.39	.72	1.0	1.52	2.77	5.2	9.54	19.2	—

In it Leverrier gave his final set of elements for the hypothetical planet (Table 4):

TABLE 4. Leverrier's Elements[5]

Epoch, January 1, 1847

Semimajor axis	36.154 a.u.
Sidereal period	217.387 years
Eccentricity	0.10761
Longitude of perihelion	284°45'
Mean longitude, January 1, 1847	318°47'
Mass	1/9300
True heliocentric longitude, January 1, 1847	326°32'
Distance from the sun	33.06 a.u.

On September 2, having revised his elements for the disturbing planet by using a new mean orbital radius a little less than what Bode's law postulated, Adams wrote again to Airy. Unfortunately Airy was in Germany and Challis, to whom Adams's new elements (Table 5) would have been most valuable, did not get to see the letter to Airy.

TABLE 5. Adams's Revised Elements

Epoch, October 1, 1846

Mean longitude, October 1, 1846	323°2'
Longitude of perihelion	299°11'
Eccentricity	0.12062
Mass	0.00015003

In France, Leverrier had had even less success in persuading French astronomers to search for his predicted planet and by the middle of September his impatience overcame his patriotism. On September 18 Leverrier wrote to Johann Gottfried Galle, then an assistant at the Berlin Observatory, asking him

---

[5]Leverrier, U.J.J., Sur la planete qui produit les anomalies observees dans le mouvement d'Uranus. *C.R. Acad. Sci. (Paris)*, 23, 428 & 657, 1846.

to look for the new planet, enclosing a copy of his revised set of elements (Table 4). Galle read Leverrier's letter and immediately asked the Observatory Director, Johann Franz Encke for permission to look for the planet. Encke reluctantly gave his consent and on September 23 Galle, assisted by an enthusiastic young student, Heinrich Louis d'Arrest, trained the Observatory's 23 cm. Fraunhofer refractor in the direction called for by Leverrier's elements: right ascension, 22^h46^m; declination, $-13°24'$. After Galle was unable to find any planetary-looking object, d'Arrest suggested using a star map. The two astronomers obtained the appropriate section of the Berlin Academy's new *Star Atlas* and Galle returned to the telescope, calling out the location and appearance of the stars for d'Arrest to check on the map. Almost immediately d'Arrest was unable to find on the star map an eighth magnitude star at right ascension $22^h53^m25^s.84$ that Galle called out. Excitedly, d'Arrest informed Encke and together the three astronomers observed the mysterious object. After determining its motion and measuring its disc with a micrometer the following night, the German astronomers were convinced of the planet's identity.

Encke was impressed by the closeness of Leverrier's predicted diameter of 3.3" and the actual measurements ranging from 2.7" to 3.2" and more still by the accuracy of the predicted coordinates. Leverrier's calculated geocentric longitude for September 23.5, 1846 was 324°58' less than one degree from Galle's actual observation at 325°52'45". On September 25 Galle wrote Leverrier informing him of the discovery and Encke followed with a most congratulatory note. On October 1, Leverrier informed Airy and Friedrich Georg Wilhelm Struve, Director of the Pulkovo Observatory in Russia, that the French Bureau of Longitudes had settled upon the name *Neptune* although it is certain that this was of Leverrier's choosing; the symbol would be a trident. However, less than a week later Leverrier decided he wanted the planet named after himself and asked his friend, Francois Arago, to promote this scheme to the French Academy of Sciences. Amidst the controversy generated by this unpopular action Leverrier became aware that a young Englishman named John Adams was also claiming to have predicted the new planet's position.

For on October 3 John Herschel, son of Sir William Herschel, discoverer of Uranus, and a noted astronomer himself wrote a letter to the *Athenaeum* in which he drew attention to the remarkable coincidence whereby Leverrier and Adams had complete-

ly independently and quite unaware of each other's efforts
arrived at virtually identical sets of elements for the new
planet. This set off a raging debate that became increasingly
political in nature. In French scientific circles Leverrier's
prior publication of his elements made his claim seem un-
assailable while the popular press incited public fervour with
emotional appeals to French patriotism and anti-English senti-
ment. In England Airy was under severe attack for having
failed to act upon or even to acknowledge Adams's predictions
with any sense of their importance and for having lost England
the glory of discovering another planet. In France Airy was
criticized for not having drawn attention earlier to Adams's
calculations when he was aware of Leverrier's efforts. In
defense Airy (1846) published his "Account of some circumstances historically connected with the discovery of the
planet exterior to Uranus." However, in less than a year the
controversy had subsided, the planet retained its original
name of Neptune and the achievement of both Adams and
Leverrier was duly recognized; indeed, the pair eventually
became friends for the rest of their lives.

On October 10, William Lassell, who earlier had lost an opportunity to search for the planet, reported having possibly detected a satellite to Neptune - less than three weeks after
the planet's own discovery. By the end of the year Neptune
had been observed by most European and American astronomers.

Chapter 2
The Search for Planet 'X'
1847-1927

As soon as Neptune's orbit was known with reasonable accuracy the residuals in the motion of Uranus, formerly as large as 133", were reduced to at most about 4". However, as the orbital elements of both planets became more refined, it appeared that the residuals of Uranus were once again considerably larger than could be explained through error of observation and orbital uncertainties.

This subject was taken up by Peirce (1848a) who published a comparison of his theory of Uranus with observation, to which he added similar comparisons of the theories of Adams and Leverrier (Fig. 2.1). Although it is not indicated by Peirce, Newcomb (1874) assumed that the numbers given in Fig. 2.1 are excesses of computed over observed longitudes.

In his paper Peirce presented the results of a complete computation of the general perturbations of Uranus by Neptune in longitude and radius vector without, however, providing any details of the investigation or of the methods employed. The smallness of the residuals in the last column (Fig. 2.1) show that by employing these perturbations by Neptune, and those by Jupiter and Saturn from Leverrier, a quite accurate set of tables for Uranus could have been constructed. However, such a table was not attempted for almost thirty years.

Newcomb (1874) noted that the research by Adams and Leverrier showed that the observed motions of Uranus could be attributed, at least approximately, to the action of a planet having the longitude of Neptune. He noted that Peirce had shown that the action of Neptune itself accounted for these perturbations within the limits of possible error of the observations used by Leverrier. However Newcomb commented that: "It remains to be seen whether the agreement between theory and observation still subsists when the comparatively few observations used by those investigators are reduced with the more refined data now

RESIDUAL DIFFERENCES BETWEEN THE THEORETICAL AND OBSERVED LONGITUDES OF URANUS, FROM THE THEORIES OF PEIRCE, LE VERRIER, AND ADAMS.

Year.	From Le Verrier's best orbit of Uranus from the modern observations without any external planet.	From Le Verrier's original theory with his best orbit of hypothetical planet, of which the mass is $\frac{1}{9332}$.	From Adams's original theory with his second hypothetical planet of which mass is $\frac{1}{6666}$.	From Peirce's theory of Neptune adopting for its mass		
				That of Struve from his own observations of the satellite $\frac{1}{14195}$.	That deduced by Peirce from Bond's & Lassel's observations combined $\frac{1}{18780}$.	That deduced by Peirce from Bond's observations of Lassel's satellite $\frac{1}{19380}$.
1690	+289.0	−19.9	+50.0	−124.7	+13.0	+0.8
1715	+279.6	+5.5	−6.6	−99.6	+10.0	+8.7
1756	+230.9	+4.0	−4.0	−102.4	−12.7	+4.0
1769	+123.3	+3.7	+1.8	−67.0	−16.0	−6.0
1782	+20.5	+2.3	0.0	−18.3	−5.6	−3.0
1787	+2.0	−1.2	−0.2	−4.7	−1.2	−0.5
1792	−7.8	−0.3	−1.1	−1.6	−0.5	−0.3
1797	−6.7	−1.0	−0.5	−3.3	+0.8	+0.3
1803	−3.4	−0.8	+1.6	−3.2	+1.2	+0.8
1808	+3.8	+0.8	0.0	−1.3	+0.6	+0.4
1813	+4.5	+0.9	+1.0	+2.3	+1.1	−0.3
1819	+3.8	−0.4	+2.2	+0.9	−0.7	+1.0
1824	−7.6	−5.4	−1.7	−1.6	−1.9	−2.0
1829	−7.8	−2.2	−2.0	+2.5	+1.3	+0.8
1835	+4.5	−0.8	+1.2	+3.9	+2.4	+2.0
1840	+0.7	+2.2	+1.3	−1.3	−1.3	−1.1
1845	+6.5	−0.3	.....	−2.8	−1.2	−0.9

Fig. 2.1. Comparison of the theories of Peirce, Leverrier, and Adams[1]

[1] Newcomb, S., An investigation of the orbit of Uranus. *Smithsonian Contr. Knowledge*, 19, No. 262, 1874. By permission of the Smithsonian Institution.

at our disposal, and when the great mass of additional observations made both before and since the date of Leverrier's researches are included."

Newcomb set out to develop a new theory for Uranus by considering the perturbations of Uranus by Jupiter, Saturn, and Neptune. He then reduced the ancient observations and those from 1781 to 1872 and compared them to his provisional theory. After applying corrections to the equations of condition Newcomb was left with a table of residuals, or outstanding excesses of the observed longitudes over theory. He said that even a glance at the residuals showed that their probable value is considerably greater than the probable error attributed to the equations of condition and that during certain periods they are of a systematic character. During the years 1748 to 1753 the observations showed a decided positive correction to the theory larger than what was reasonable to expect. About 1800 the correction became negative for about twenty years with an average value of about 1". In 1821 it suddenly became positive until 1833 after which it exhibited an irregular pattern. Newcomb pointed out that the residuals were much greater than the purely accidental residuals resulting from the theory of least squares and listed three main types of possible causes of the errors.

1. Systematic errors of observation of which the two main causes are a deviation of the line of collimination of the instrument from a true great circle, or from any peculiarity of the observer which leads him to record the transit of Uranus earlier or later than a fixed star.

2. Errors in the theory under comparison, which may arise from errors in the computations, from the omission of the terms of the second order produced by Neptune, from the adoption of an erroneous mass of Saturn, or from the attraction of an unknown planet.

3. Errors in the various reductions by which theory and observation are compared. In the method adopted for this comparison a number of small uncertainties incident to the imperfections of the older data of reduction necessarily creep in.

In discussing the probability of these different sources of

error Newcomb commented that a trans-Neptunian planet large enough to produce a sensible deviation of the orbit of Uranus from an ellipse in the course of a century would be too large to have escaped detection. After corrections Newcomb found the following mean elements:

TABLE 6. Newcomb's Mean Elements of Uranus

Epoch, January 0, 1850, Greenwich Mean Noon[2]

Longitude of the perihelion	170°38'48".7+8698".μ
Mean longitude at epoch	29°12'43".73+2811".4μ
Longitude of the node	73°14'37".6+29".6μ
Inclination of the orbit	0°46'20".92+0".38μ
Eccentricity	.0463592-5236μ
Eccentricity in seconds	9562".27-108".0μ
Mean motion	15424.797-0".838μ
Log mean distance (uncorrected)	1.2829251+179μ
The same corrected	1.2831223+179μ
True mass of Neptune	$\frac{1 + \mu}{19700}$

Newcomb provided tables, based on the above elements and his theory, for calculating the position of Uranus for any year and date from 1000 to 2200 A.D.

In his book *Astronomie Populaire*, published in November, 1879, the French astronomer Camille Flammarion deduced the existence of a trans-Neptunian planet from the aphelia of the comets of 1862III, 1532, and 1661. Flammarion (1884) later published a paper on the same subject. He predicted a large planet with an orbital radius of about 43 a.u. and a period of 330 years.

Forbes (1880a) noted that "it has long been known that the greatest distances (aphelion distances) to which comets recede from the sun are grouped into classes." He pointed out the large class of comets with aphelia clustered around Jupiter's

---

[2]Newcomb, S., An investigation of the orbit of Uranus. *Smithsonian Contr. Knowledge*, 19, No. 262, 1874. By permission of the Smithsonian Institution.

orbital radius and another class with aphelia equal to Neptune's orbit. He then said that he had noticed other groupings of cometary aphelia at 100 a.u. and 300 a.u. and that "there could be no longer a doubt that two planets exist beyond the orbit of Neptune, one about 100 times, the other about 300 times the distance of the earth from the sun, with periods of revolution of about 1000 and 5000 years respectively." Forbes summarized his evidence for the 100 a.u. group of aphelia in a table of data (Table 7):

TABLE 7. Forbes's Table of Cometary Apehelia[3]

	Date	Aphelion Distance (a.u.)	Date of Aphelion (A.D.)	Calculated Period (Years)	L.
I.	1840, iv.	96.7	1668	350	313°
II.	1843, i.	100.0	1655	376	225°
III.	1846	108.2	....	...	...
IV.	1861, i.	110.3	1654	413	139°
V.	1793, ii.	111.0	....	...	...
VI.	1861, ii.	111.2	....	...	...
VII.	1855, ii.	124.2	1608	493	192°

Forbes observed that it was impossible for the hypothetical planet to have in one revolution, either by direct or retrograde motion, passed each of the aphelion positions listed above at the aphelion times. It was a matter of considerable effort to determine whether this could have happened in even two or three revolutions. He tried a large number of hypotheses before finding one agreeing well with facts. He presented elements for a hypothetical planet at a mean distance of 98 a.u. Forbes concluded by commenting, in support of his hypothesis, that he had computed the position of Neptune by its influence on comets to within 2°, although previously ignorant of its actual position.

---

[3]Forbes, G., On comets. *Proc. R. Soc. Edinburgh,* 10, 426, 1880. By permission of the Royal Society of Edinburgh.

Forbes (1880b) presented a second paper providing further details of the elements of his hypothetical planet. He found by calculation that the effect of Neptune upon Uranus was about ten times that of his new planet. But the latter was three times farther from the sun than Neptune and hence its effect was inversely proportional to the cube of its distance. Moreover, in the case of Neptune the period during which the perturbation was greatest was 75 years, while in the case of the new planet it was only 45 years. Accordingly Forbes calculated the mass of the new planet to be 1964 reciprocal solar masses or ten times Newcomb's (1874) value for the mass of Neptune, second only to Jupiter in the solar system. Forbes reported having received a letter from Mr. D.P. Todd of the Nautical Almanac Office in Washington, D.C. Todd revealed in this letter that Forbes's position for the interior (98 a.u.) of his two planets differed by only four degrees from the position Todd had provisionally calculated for a planet responsible for the residuals of longitude of Uranus and Neptune. He went on to describe the unsuccessful search he had made for his planet with the 65-cm. refractor of the U.S. Naval Observatory. Forbes was encouraged to distribute about a hundred copies of his paper to observatories which resulted in quite a number of these taking up searches for the planet. Roberts (1892) took two sets of 18 plates with each plate covering more than four square degrees, the exposure being 90 minutes long. A close examination of the plates by Roberts showed that "no planet of greater brightness than a star of the fifteenth magnitude exists in the sky area herein indicated."

In addition to his letter to Forbes, Todd (1880) published a full account of his planetary search. He had in mind a planet at a mean distance of 52 a.u., with a diameter of 80,000 kilometers, subtending an angle of 2".1. On thirty clear, moonless nights between November 3, 1877 and March 5, 1878, he searched a double strip one degree wide along either side of the invariable plane from longitude 146°.8 to longitude 186°.1, employing magnifying powers of 400-600 times and looking for an object with a perceptible disk. After checking out many suspicious objects, Todd was confident that no such planet existed, at least where he had searched.

Gaillot (1909) proposed two trans-Neptunian planets, one with a mass of five times that of the earth and a mean distance of 44 a.u., and the second with 24 earth masses and a mean distance of 66 a.u., as possible causes of the perturbations in the motions of Uranus and Neptune. The more massive planet was predicted to have a longitude, adjusted to epoch 1930.0, of 128°.5; the less massive planet was situated at longitude 308°.4 (epoch 1930.0). Lau (1914) predicted two planets of 24 and 48 earth masses and mean distances of 46.5 and 71.8 a.u., respectively (see Crommelin, 1931).

Pickering (1909) used an elaboration of the graphical procedure described by Sir John Herschel (1849) to predict the existence of a trans-Neptunian planet, which he called planet O, from its perturbations of Neptune and to a lesser extent, Saturn and Uranus. He drew graphs of the displacements produced when he plotted the residuals in longitude of the three outer planets, correcting them as well as he could for errors in period, eccentricity and apse line. Then he looked for the characteristic signs of a conjunction with the unknown planet. These take the form of a noticeable tendency of the residuals towards positive values followed by a pronounced change to negative values as the known planet is first pulled ahead of its computed place by the unknown planet then retarded from its normal motion by the unknown. Pickering thus concluded that Neptune had been in conjunction with the unknown about the beginning of this century and that a conjunction between the unknown and Uranus had occurred in about 1850.

Pickering considered his empirical method much simpler and quicker than the analytical method used by Adams and Leverrier and subsequently Lowell. Although his method did not provide the orbital elements of the unknown, only its position, Pickering considered the important thing was to find the planet, and that the orbit was of no use or consequence other than to correct the heliocentric longitude from the date of maximum perturbation. Pickering (1928a) later predicted that planet O would be small and that its orbit would be eccentric and would cross that of Neptune meaning that planet O would sometimes be closer to the sun than Neptune (Fig. 2.2). This would be unique in the solar system.

Pickering also noted the curious fact that although planet O moved in a direct sense about the sun, to an imaginary inhabitant of Neptune it would appear to move in a retrograde orbit like Neptune's satellite Triton.

About the same time as Pickering's work, Gaillot (1910) published an analytical reduction of perturbations of Uranus by an unknown planet. Gaillot incorporated several observations between 1873 and 1882 Pickering did not have access to and also substituted more modern values for the masses of the larger planets than those used by Leverrier in 1877. The results of Leverrier, Gaillot and Pickering are compared in Fig. 2.3.

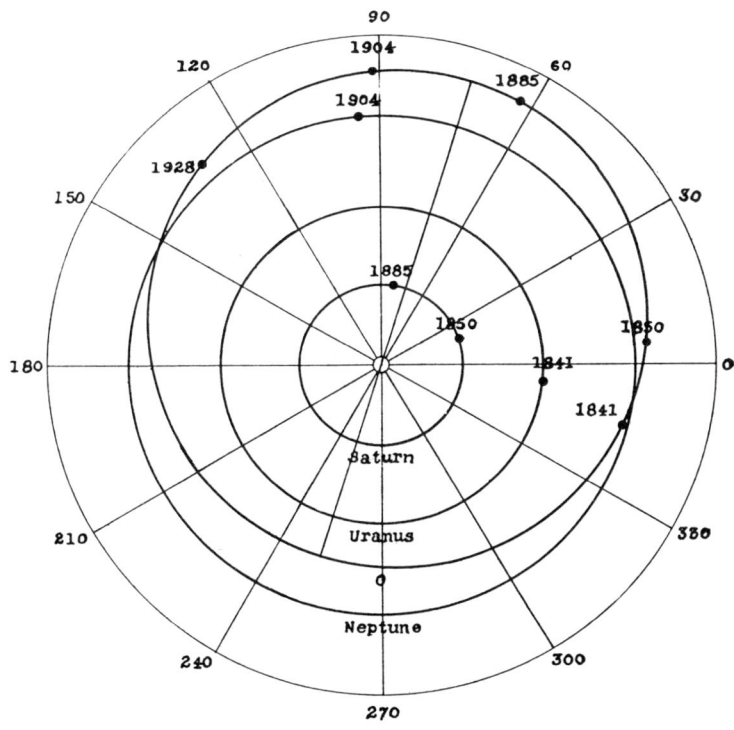

Fig. 2.2. Elliptical orbit of planet O^4

---

[4] Pickering, W.H., The next planet beyond Neptune. *Pop. Astron.*, 36, 143, 1928.

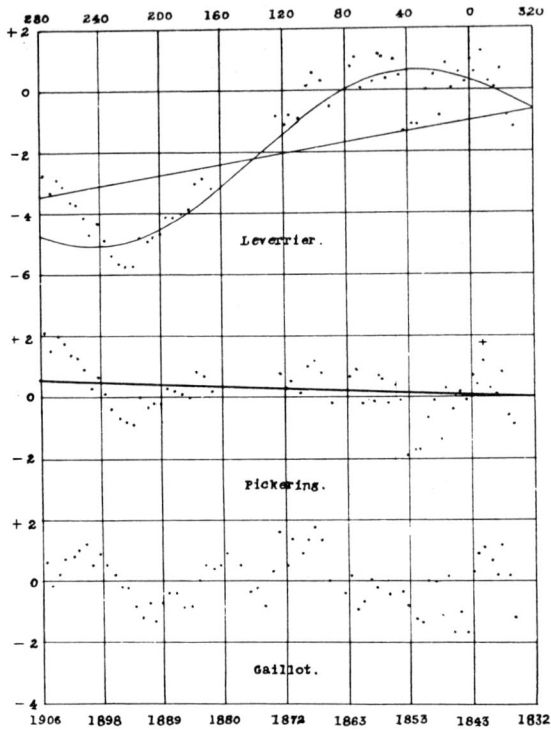

Fig. 2.3. Accepted and revised orbits of Uranus perturbed by planet O^5

(Pickering, 1928a)

The name most intimately associated with Pluto is that of Percival Lowell. For a more detailed account of this gifted amateur astronomer, who built an observatory to rank very high among the observatories of the world and who not only devoted all of his time to astronomy but also engaged a full staff of professional astronomers, the reader is referred to Abbot Lowell's (1935) excellent biography.

---

[5] Pickering, W.H., The next planet beyond Neptune. *Pop. Astron.*, 36, 143, 1928.

In his *Memoir on a Trans-Neptunian Planet*, Lowell (1915) gave, in a little more than a hundred pages and with little mention of the enormous labor involved, the results of his decade-long research on the theoretical evidence of a planet beyond Neptune. Lowell chose the name "planet X" for this unknown member of the solar system. As noted by Putnam and Slipher (1932) the best way to introduce the work of Lowell is to quote from the beginning of his classical *Memoir:*

> Ever since celestial mechanics in the skillful hands of Leverrier and Adams led to the world-amazed discovery of Neptune, a belief has existed begotten of that success that still other planets lay beyond, only waiting to be found. Leverrier, himself, with the far-sight of genius, was firmly of this view, though unfortunately oversanguine of the happy date of its demonstration. In consequence, since his time, many attempts have been made to indicate the position of one or more of these unknowns, attempts for the most part, of no scientific value because not founded on rigorous mathematical investigation. For so complicated is the problem that all elementary means of dealing with it lead only to error. The sole road to any hope of capture lies through the methodical application of laborious analysis.
>
> Not only are all summary processes worse than useless, engendering false ideas in their conclusions, even a supposed proximate analysis proves not to be approximate in its results. Thus the simplifying of the problem by the assumption of a circular orbit for the unknown originally suggested by Tisserand and worked out in the present case with all due reserve by M. Gaillot, and more or less likewise by M. Lau, betrays intrinsic evidence of inadequacy....
>
> 2. To any real solution, the problem must be attacked analytically with all the rigor possible. Before entering upon such an investigation, it is well to state the problem generally, showing the data upon which it rests, and the limitations to which it is necessarily subject. For lack of appreciation of these points has led to mistaken ideas of what is or is not possible.
>
> 3. The theory of a planet can not in the nature of things be exact; and this for three reasons:

(1) The observations on which it is founded are necessarily more or less in error;
(2) The theory, itself, may be more or less imperfect;
(3) An unknown body may be acting of which perforce no account has been taken....[6]

Lowell adopted the old analytical method invented by Adams and Leverrier and made use exclusively of the perturbations of the planet Uranus. He could not use Neptune with his method because its observed orbit was only half completed and, as a result, the uncertainties in its elements were as large as the perturbations being sought. Lowell modified the old analytical method in several ways. He made solutions for various distances for planet X rather than employing Bode's law as had Leverrier in the case of Neptune. He used least-squares reductions almost exclusively and limited his development to the perturbations in longitude. Lowell used a second order solution because of the large eccentricity of Pluto's orbit. He used the percentage reduction of the residuals of Uranus as a measure of the fit of his solution. For the computation of the longitude of planet X, Lowell made the initial assumption that it travelled in the same plane as Uranus; leaving, therefore, the following five elements to consider:

a', the mean distance (or n, the mean motion)
ε', the mean longitude at the origin of the time
e', the eccentricity
ω', the longitude of perihelion
m', the mass.

It was also necessary to include in the solution corrections Δn, $\Delta \varepsilon$, Δe, and $\Delta \omega$ to the corresponding elements of Uranus, making a total of nine unknowns. The general procedure Lowell used was to assume values for a' and ε' and to solve for the seven remaining quantities by the method of least-squares; repetition of this process many times, with different values of a' and ε', made it possible to eventually find values for all nine unknowns that resulted in the largest percentage reduction in the residuals of Uranus.

The observation data used by Lowell consisted of the devi-

---

[6] Lowell, P., Memoir on a trans-Neptunian planet. *Mem. Lowell Observ.*, I, 1, 1915. By permission of the Lowell Observatory.

ations in the observed longitude of Uranus from those given by Gaillot's (1910) theory; these were grouped into 37 observation equations covering the period from 1690 to 1910. Lowell first made a preliminary partial solution, which he designated (H_{14}), in which 24 observation equations, covering the period of 1750 to 1903, were used. In this solution 25 perturbation terms, involving e', e, and e^2, were taken into account; a' was set at 47.5 a.u., and solutions were made for 30 values of ε' from 0° to 345°. The main result of this partial solution was to locate ε' approximately, so that in the complete solution (repeated for other values of a') it was not necessary to go all around the circle, but only to cover two parts of it. The complete solution, designated (H_{14}), was carried out for values of a' ranging from 40.5 a.u. to 51.25 a.u. Lowell's final results are summarized in the last two pages of his *Memoir* (Fig. 2.5 and Fig. 2.6).

Thus Lowell arrived at a double solution for the position of planet X. The situation was analogous to trying to predict the moon's position from observations of the lunar tides. Lowell's two best solutions, with $\varepsilon' = 0°$ and 180°, reduced the outstanding residuals of Uranus by 99 percent and 90 percent, respectively; this small difference was insufficient, in Lowell's opinion, to distinguish between the two.

Although Lowell published his final results in 1915, he had begun his researches much earlier. Indeed, his earliest calculations were so encouraging that the preliminary steps on the search for planet X were taken under his direction in 1905. Lowell realized that photography offered the only means of finding such a faint planet among the numerous stars. After having some preliminary test exposures made in 1905, he purchased a 12.5-cm photographic objective from the John Brashear Company. With this 12.5-cm camera E.C. Slipher secured about two hundred plates in 1906 and 1907 and K.P. Williams took another fifty in the summer of 1907. This series of photographs, recording stars below the 16th magnitude with three-hour exposures, covered the entire circuit of the sky centered along the invariable plane, duplicate plates being made every 5°.

Lowell at first examined these plates with a hand magnifier but soon found this to be inadequate (E.C. Slipher later found two faint images of planet X - Pluto - that Lowell had overlooked on the 1905 plates). He then had a Hartmann comparator, intended for small spectrum plates, enlarged and adapted

A TRANS-NEPTUNIAN PLANET

as further study has shown this confidence to have been misplaced; so the fine definiteness of positioning of an unknown by the bold analysis of Leverrier or Adams appears in the light of subsequent research to be only possible under certain circumstances. Analytics thought to promise the precision of a rifle and finds it must rely upon the promiscuity of a shot gun after all, though the fault lies not more in the weapon than in the uncertain bases on which it rests. But to learn of the general solution and the limitations of a problem is really as instructive and important as if it permitted specifically of exact prediction. For that, too, means advance.

SUMMARY

69. This investigation establishes the following:

1. By the most rigorous method, that of least squares throughout, taking the perturbative action through the first powers of the eccentricities, the outstanding squares of the residuals from 1750 to 1903 have been reduced 71% by the admission of an outside perturbing body.

2. The inclusion of further terms yielded solutions in accordance with the first.

3. Solutions taking the years 1690-1715 also into account agreed substantially with those from the years 1750-1903.

4. So did those in which the additional years to 1910 were considered.

5. The second part of the investigation, in which the solutions were made for the second powers of the eccentricities as well, gave conformable results.

6. When the probable errors of observation were reckoned, the outstanding squares of the residuals of theory excluding an outside planet proved to have been reduced by its admission from 90% to 100% nearly, the solutions seeming to confirm one another as follows:

for e' around $180°$ and for e' around $0°$

24 obs. eqs. 99½%
25 "   " 91%
27 "   " 80½%
37 "   " 88%

99½%
99½%

A TRANS-NEPTUNIAN PLANET

7. Though this would indicate an absolute solution of the problem, it must be remembered that the actual as against the probable errors of observation might decidedly alter the result; and so might the terms above the squares in e and e' necessarily left out of account.

8. The investigation disclosed two possible solutions in each case, one with e' around $0°$, one with it around $180°$; and that this duality of possible place would necessarily always be the case.

9. On the whole, the best solutions for the two gave:

e' around $0°$ e' around $180°$
$e' = 22°.1$ $t' = 205°.0$
$a' = 43.0$ $a' = 44.7$
$m' = 1.00$ $m' = 1.14$
$e' = .202$ $e' = .195$
$\bar{\omega}' = 203°.8$ $\bar{\omega}' = 19°.6$

hel. long. July 0, 1914 $84°.0$ $262°.8$

the unit of $m' = \frac{1}{50000}$ the mass of the Sun.

10. It indicates for the unknown a mass between *Neptune's* and the *Earth's*; a visibility of the 12-13 magnitude according to albedo; and a disk of more than 1″ in diameter.

11. From the analogy of the other members of the solar family, in which eccentricity and inclination are usually correlated, the inclination of its orbit to the plane of the ecliptic should be about 10°. This renders it more difficult to find.

12. Investigations on the perturbation in latitude yielded no trustworthy results. This is probably because the eccentricity e' as well as the planet's other elements enter as data into the latitude observation equations.

13. The perturbative function is not discontinuous at the commensurability of period points, a fact hitherto in doubt.

14. That when an unknown is so far removed relatively from the planet it perturbs, precise prediction of its place does not seem to be possible. A general direction alone is predicable.

Fig. 2.4. (opposite page)

The last two pages of Percival Lowell's historic *Memoir*.[7]

---

[7] Lowell, P., Memoir on a trans-Neptunian planet. *Mem. Lowell Observ.*, I, 1 (1915). Courtesy of Lowell Observatory.

to examine the sky plates. Later still he procured a Zeiss Blink comparator.

Lowell next had an extensive series of plates made with the 105-cm reflector by Slipher and C.O. Lampland. Although the light-power of this large aperature instrument greatly reduced the required exposure times, the individual plates covered such a small field of view that thousands were needed to cover even a moderate area of sky. It was soon apparent that what was needed was a wide-field, photographic telescope of considerable light-power. Unfortunately the advent of World War I delayed the acquisition of this instrument and Percival Lowell died in 1916 before either his planet X or the instrument needed to find it were located. In 1925, V.M. Slipher suggested to Mr. Guy Lowell, then Trustee of the Lowell Observatory, that a set of 32.5-cm glass disks be purchased for the photographic instrument. Guy Lowell obtained the glass blanks just before his death in 1927 and shortly thereafter the necessary money to complete the instrument and its mounting was generously provided by Percival Lowell's brother, A. Lawrence Lowell, President of Harvard University.

Pickering (1919) later used the residuals in longitude of Uranus and Neptune to predict the position of planet O for epoch 1920.0. He also used Neptune's residuals in latitude to predict planet O's inclination and node. Pickering's (1909, 1919) prediction and that of Lowell (1915) prompted M.L. Humason at Mt. Wilson Observatory to take a series of plates in December, 1919 and January, 1920 with the 25-cm Cooke triplet telescope. The plates were exposed for two hours and stars down to magnitude 17 were recorded. The area examined critically on the plates was confined to the region within 2° of the ecliptic and no images of either Pickering's planet O or Lowell's planet X were found. This was unfortunate because it seems to have led Pickering to distrust his own prediction and to believe that no such single planet existed, for he later (1928d, 1931c) published work based on a plurality of unknown planets. Certainly in 1930 he did not at first believe that the object discovered at the Lowell Observatory could be a planet. Ironically, as soon as Pluto was discovered and its orbit accurately determined, the 1919 plates were re-examined by S.B. Nicholson and four very faint images of Pluto were detected.

Crommelin (1931c) summarized the elements of the hypothetical planets of Lowell, Pickering, Gaillot, Lau, Todd and Forbes in a table (Table 8).

TABLE 8. A Comparison of the Elements of Pluto and Predicted Trans-Neptunian Planets[8]

	P. Lowell, 1915.	W. H. Pickering, 1909.	W. H. Pickering, 1919.	W. H. Pickering, 1928.	Gaillot, 1909.	Actual Elements of Pluto.	P. Lowell, 1915.	Gaillot, 1909.	Lau.	Todd, 1877.	Forbes, 1887.	Lau.	See, 1904.
Mean dist.	43.0	51.9	55.1	30.1	66	39.5	44.7	44	46.5	52.0	104.4	71.8	42.2
Period, years	282	373.5	409.1	164.8	536.1	248	299	292	317	375	1066	608	275
e	.202	..	.31	.195	..	.248	.195	..	..	..	..	..	..
Long. perihelion	204°.9	..	280°.1	252°.5	..	223°.4	20°.7	..	..	..	..	..	..
Date perihelion	1991.2	..	1720.0	1973.8	..	1989.8	1994.9	..	..	..	..	..	..
Ω	..	..	100°	180°	..	109°.4	..	..	..	104°	..	..	..
i	10°	..	15°	..	..	17°.1	..	..	..	1°.4	..	..	..
Long. 1930.0	102°.7	135°.1	102°.6	135°	128°.5	108°.5	279°.0	308°.4	308°.9	221°.6	191°	173°.1	233°
Magnitude	12 to 13	13.4	15	{12.2 v / 13.5 p}	..	15	12 to 13	..	..	13	..	..	..
Mass. (Earth=1)	6.7	2	2	0.75	24	..	7.6	5	9	..	..	48	..

v, p denote visual and photographic magnitudes.

[8] Crommelin, A.C., The Discovery of Pluto. *Mon. Not. R. Astron. Soc.*, 91, 380 (1931c). By permission of the Royal Astronomical Society.

Chapter 3
The Discovery of Pluto
1928-1930

Pickering (1928c) stated that "the most productive orbit in the solar system today, by means of which one may search for unknown planets is undoubtedly that of Uranus." In his graph of the perturbations of Uranus, Pickering found more evidence for his planet O and also new evidence for another planet, P, at a distance from the sun of 61.7 a.u. and a period of 485 years, located at a longitude of 294°. He also found evidence for a planet S, located at longitude 344°, which he had earlier postulated on on the basis of cometary aphelia data. In his next paper, Pickering (1928d) noted that little more information regarding planets P and S could be extracted from the residuals of Uranus. However, he pointed out that another source of information for determining the elements of the orbit of planet P was available from the study of cometary aphelia. On the basis of his investigation of this information, Pickering found a revised period of 556.6 years and a distance from the sun of 67.7 a.u. for planet P. He gave its position for epoch 1920.0 as $\alpha = 20^h 19^m$ and $\delta = -51°.5$. He predicted its visual magnitude to be 11.0 and therefore confidently stated that it would not be difficult to detect. He attributed the fact that planet P had not yet been found accidently to its far southern declination.

With regard to planet S, Pickering observed that because there were only four associated comets, little reliance could be placed on his predictions of the inclination and the node. However he gave a tentative position for epoch 1920.0 of $\alpha = 22^h 04^m$ and $\delta = +3°.4$. Concerning planet O, Pickering predicted its orbit would be highly inclined as eccentric orbits often are. He predicted a visual magnitude of 12.2 and a photographic magnitude of 13.5 for planet O. As mentioned in the preceding chapter, the failure of M.L. Humason's 1919 photographic search at Mt. Wilson to find Pickering's planet O perhaps led Pickering to distrust his own prediction and to believe that no single such planet existed. When the discovery of Pluto was first announced, in March, 1930, Pickering (1930c) seemed convinced that it was a comet, rather than a true planet. He lamented that if it had been announced when discovered (in February, 1930), as was customary, it would have been possible to know its orbit by then.

The completed 32.5-cm objective lens for the Lowell Observatory's planetary search camera arrived in Flagstaff in early 1929. As soon as the lens was placed in its mounting the preliminary adjustments and tests were rushed to completion. After an initial trial series of plates of Lowell's predicted favourable region in Gemini was completed, Mr. Clyde W. Tombaugh who had recently joined the observatory staff was appointed to carry out the search for planet X. It was found that the field of good images was large enough to use 35 x 42-cm plates. The focal length of the 32.5-cm objective was 169 cm, giving a scale of 122"/mm or 2.95 cm per degree. Thus the plates covered an area of the sky nearly 12° by 14°. Special plate holders were designed having special bearings for the plates over which thumb screws sprang the plates into adjustment with the focal surface of the large lens.

The observation program was started on April 1, 1929 with plates being taken of the Gemini region. It was soon found that photography near the opposition point was essential if the task of distinguishing between distant planets and asteroids near their stationary points was not to be made nearly impossible. Near the opposition point the angular motions of the asteroids were much greater and their short trailed images during the one-hour exposures revealed their true identity at a glance.

By September, 1929, the photographic work had caught up with the opposition point on the ecliptic. Blink examination of the plates then proceeded with increased care and thoroughness. No matter how carefully the plates were processed and handled a number of defects, ranging from particles of dirt or other foreign material to metallic silver deposits in the emulsion, could not be avoided. Only negatives were examined; positive copies would have contained many more defects.

The first blink examination of opposition plates began with those of the constellation of Aquarius in September and continued month by month, eastward, through Pisces, Aries, and Taurus. By then the search was well into the starfields of the Milky Way and by the time the examination of the Taurus plates was completed, the Gemini plates had been taken. Tombaugh first placed a pair of δ Geminorum plates on the Zeiss blink comparator. Three plates were normally taken of each region at two day intervals. However the interval between the date of the first δ Geminorum plate, January 21, 1930, and the third plate, January 29, was four days longer than usual. When a quarter of the δ Geminorum plate pair had been examined, on February 18, Tombaugh detected a "planet suspect". The images were 3.5 mm apart - a shift that indicated the object lay beyond the orbit of Neptune. The images were also found on the 12.5-cm Cogshall plates taken concurrently on January 21, 23, and 29.

Tombaugh was able to secure a fourth plate on February 19th and the image was again found in the predicted place. There were early fears that the object would speed up, and turn out to be only an exceptional asteroidal or cometary body, as was subsequently suggested by Pickering and Jackson; but succeeding plates always showed the object in its predicted place. On February 20th, the observatory staff examined the object visually with the 60-cm refractor to see whether it exhibited a disk or not. It did not and because Lowell had predicted a 1" diameter there was now some suspicion that the main body had yet to be found. A few nights later, C.O. Lampland, using the 105-cm reflector, compared the color of the object photographically with that of Neptune. His comparison confirmed the visual impression that it was distinctly yellowish rather than bluish like Neptune. One-hour exposure plates taken with the 105-cm instrument failed to reveal any satellites; consequently there was no immediate way to determine the object's mass. E.C. Slipher continued to examine the new planet visually with the 60-cm refractor on nights of excellent seeing conditions. Meanwhile Lampland started a long series of short-exposure plates for astrometric purposes with the 105-cm reflector, securing a plate in every possible month until his death in 1951 (Fig. 3.1).

Finally, on the night of March 12, 1930, a message was telegraphed to the Lowell Observatory trustee, R.L. Putnam, for forwarding to Harvard College Observatory for distribution, the next day, to observatories and astronomers. The historic telegram read:

> Systematic search begun years ago supplementing Lowell's investigations for Trans-Neptunian planet

has revealed object which since seven weeks had in
rate of motion and path consistently conformed to
Trans-Neptunian body at approximate distance he
assigned. Fifteenth magnitude. Position March
twelve days three hours GMT was seven seconds of
time West from Delta Geminorum, agreeing with
Lowell's predicted longitude.[1]

The announcement of the discovery of the new planet was thus made on March 13th, which by a double coincidence was the anniversary of Lowell's birthday and also the 149th anniversary of the discovery of Uranus by Sir William Herschel. The news of the discovery was telegraphed by Harlow Shapely to the Royal Astronomical Society where it was read at a meeting the following day. The Society's official reply, prepared by its Foreign Secretary on behalf of the Council, was then read:

President, Council, and Fellows of the Royal Astronomical
Society in meeting assembled send to Lowell Observatory
their hearty congratulations on the great discovery of
Trans-Neptunian planet. - TURNER, Foreign Secretary.[2]

The reaction of the meeting was duly recorded in the minutes by the recording Secretary as "applause".

The announcement was also made public on the morning of March 13th by Percival Lowell's widow, Mrs. Constance S. Lowell, during the presentation of a scholarship prise, given annually in Lowell's memory, at the Arizona Teachers College.

The Lowell Observatory staff were overwhelmed by the "tremendous, instant, and persistent" demand from the press media for information. The number of telegrams, cablegrams, and letters received at the observatory, as well as the large coverage given the discovery by newspapers and magazines prompted the staff to determine as soon as possible a preliminary orbit for the planet. A circular giving the preliminary orbit and some further results of the observational work was communicated to Harlow Shapely at the Harvard College Observatory. Shapely

---

[1] Slipher, V.M., A trans-Neptunian planet. *Pop. Astron.*, 38, 187 (1930).
[2] From: Meeting of the Royal Astronomical Society, Friday, 1930 March 14. *Observatory*, 53, 97 (1930). By permission of the Royal Astronomical Society.

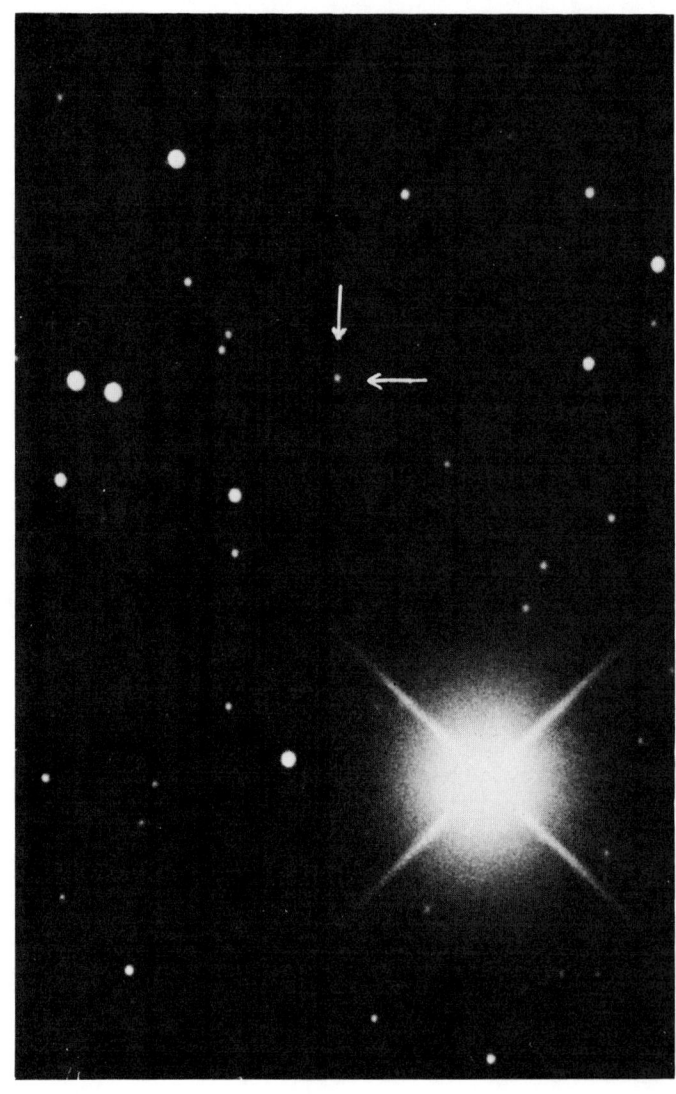

Fig. 3.1. R.A.S. No. 416. Pluto. Lowell Obs. 42-in. Refl. *F.L.* 220 in. 1930 March 2d. 4h. 56m. ; 5d. 3h. 44m. Lampland.[3]

---

[3]Courtesy of the Lowell Observatory.

(1930b) then distributed the Lowell data as *Harvard College Observatory Announcement Card*, No. 121:

> Preliminary orbit Planet X has been computed by Lowell Observatory staff with collaboration of Dr. John A. Miller, director of the Sproul Observatory, using positions January 23, February 23, and March 23 determined from Lowell plates by Lampland and yielded following elements referred to mean equinox 1930.0.
>
> | | |
> |---|---|
> | Node | 109°21' |
> | Inclination | 17°21' |
> | Log, semi-major axis | 2.3359 |
> | Longitude perihelion | 12°52' |
> | Eccentricity | 0.909 |
> | Mean daily motion | 1".112 |
> | Mean anomaly 1930.0 | 3°20'47" |
> | Distance from Sun | 41.3 |
>
> Position computed from orbit for March 30 checked observed position within small fraction of one second of arc in both coordinates. However, because of very short arc available and object's extremely small latitudes, considerable revision of some elements, especially eccentricity, is not expected. Our first preliminary circular orbit computed from positions January 23 and March 23 had given node and inclination agreeing with these of elliptic orbit. Further details will appear promptly in *Lowell Circular*. V.M. Slipher.[4]

R.L. Putnam, who had forwarded the discovery note from Lowell Observatory to the Harvard College Observatory, had suggested that the credit be given to V.M. Slipher, director of the Lowell Observatory. However Harlow Shapely felt that credit for the discovery belonged to Percival Lowell who had mathematically predicted the new planet and "next to the administrators of the Lowell Observatory and to the observers following their nightly routine." Shapely suggested the name Cronos for the new planet. Frank Schlesinger, director of the Yale Observatory predicted that other major planets would be added to our solar system although they would be increasingly difficult to

---

[4]Shapely, H., Trans-Neptunian planet. *Pop. Astron.*, 38, 206 (1930b).

detect owing to their faintness. Captain Charles Freeman of
the U.S. Naval Observatory said that the scientific community
was "very much gratified and very much thrilled" by the discovery. He said that the U.S.N. Observatory would attempt to
observe the planet although, at 15th magnitude, it would be
elusive especially since its position had not yet been given
with any degree of accuracy. Otto Struve of Yerkes Observatory hailed the discovery as "one of the greatest in the history of astronomy." J.Q. Steward of the Department of Astronomy at Princeton University called the discovery "an important
and interesting contribution to the knowledge of the solar
system." Sir James Jeans made the rather prophetic prediction
that the new planet "may prove to have a single relatively
large satellite, like Neptune and the earth."

Pickering (1930a) commented that since the discovery of the
new planet had excited such great interest it was desirable
that a name should be found for it as soon as possible. He
suggested that it should be called "Pluto" after the Greek god
of Darkness, who was also able at times, when he so desired,
to render himself invisible. Pickering noted that the new
planet when at aphelion received less than 0.002 times as much
light from the sun as we do, and appears to us then only 1/19th
as bright as when at perihelion.

Two days after she had made the first public announcement of
the new planet's discovery, Mrs. Lowell suggested that it be
named "Percival" after her late husband. Half a week later,
Mrs. Lowell changed her mind and said she preferred "Lowell"
for the name of the new planet. However Captain Freeman later
suggested Percival Lowell would have vigorously opposed the
use of his name for the new planet as had Herschel similar
attempts to name Uranus after himself. Freeman proposed instead "Minerva", noting that "Cronos" was identical with Saturn
while "Pluto" might be undesirable due to its association with
Satan. Slipher (1930b) reported that of the many names
suggested for the new planet, "Minerva" and "Pluto" were very
popular. But, as he noted, Minerva had already been used for
the name of an asteroid. However, Pluto seemed appropriate
and the Lowell Observatory proposed this name to the American
Astronomical Society and the Royal Astronomical Society.
Slipher noted that, as far as was known, the first person to
suggest the name Pluto was Miss Venetia Burney, aged 11, of
Oxford, England. As a fitting symbol, the Lowell Observatory
suggested ♇, easily remembered because they are the first two
letters of the name. The new symbol was also, not coincident-

ally, a monogram of Percival Lowell's initials.

About a week after Pluto's discovery, J. Jackson of the Royal Observatory at Greenwich who had earlier expressed the view that "although we have little details as yet, the fact that the observations extend over seven weeks should make sure that the object is really a very distant planet" suddenly changed his opinion. Now, Jackson said, he wasn't so sure that Pluto was a planet because: first of all, it wasn't found exactly where Lowell predicted it would be found; nor did it have the predicted brightness. Also he thought it odd that Lowell Observatory had released so few details. In reply to Jackson's new skepticism, Shapely (1930a) reported that Harvard College Observatory had received additional observations from Yerkes Observatory. G. Struve of Potsdam Observatory reported the visual and photographic observation of Pluto with the 65-cm refractor; it was also detected at the Heidelberg Observatory. Bianchi of the Brera Observatory near Milan, Italy reported detecting the new planet on plates taken on March 16th. Captain Freeman of the U.S. Naval Observatory reported that Pluto had been photographed with the 25-cm photographic telescope and seen visually with the 65-cm refractor. J.S. Paraskevopoulos (1930) at the Boyden (South Africa) Station of the Harvard Observatory reported obtaining photographs of the new planet with the 60-cm Bruce refractor on March 18, 19 and 27 and April 16 and 18.

Esclangon (1930a) reported that micrometer measurements of photographs of Pluto at the Paris Observatory showed that a minimum of right ascension was passed at March 30.5, 1930. The earliest calculations gave a circular orbit with a period of 489 years. Using Laplace's method Mineur and Canavaggia at the Paris Observatory showed that Pluto's velocity was constant in speed and direction during the first 18 days of observation. Employing a parallactic method Stoyko found (1) the heliocentric coordinates of the planet referred to the ecliptic and the equinox; (2) the distance from the sun, R, 44.4 a.u.; (3) the longitude of the node, Ω, 108°55'; and (4) the inclination, i, 38°54'. Of these, Esclangon noted, only R and Ω were reliable and many months or years of observations would be needed to compute an accurate orbit. Later, using pairs of observations on either side of the minimum, Stoyko found by a modification of Lagrange's method R = 41.524 a.u., Ω = 109°16' and i = 19°51'.

Jekhowsky (1930) also calculated the heliocentric distance of

Pluto by Lagrange's method from early observations and new observations as they became available. He reported values for R of 44.219 a.u. for March 25, 1930 and 41.595 a.u. for April 2. Stoyko (1930a) calculated new elements for the orbit of Pluto using prediscovery observations made on March 16 and April 1, 1926. He found a lower limit for the eccentricity of 0.407; but he noted that its actual value was still unsettled.

Banachiewicz (1930c) noted that the observations of Pluto in March and April, 1930 were insufficient to calculate its orbit, if any three observations were employed, because of the doubt regarding the actual path of the planet. He outlined a new method in which linear equations could be deduced to give the coordinates and vectorial speed of the planet by considering a projection of the path of the planet on the earth. Banachiewicz said that this method was able to determine the nature of Pluto's orbit from observations obtained in the first half of 1930, while all other methods based on three observations would give erroneous results. Esclangon (1930d) used photographic observations of Pluto made between August and October, 1930 at the Paris, Yerkes, and Heidelberg Observatories combined with the Paris observations for January to May, 1930 to recalculate Pluto's orbit. He found that the planet, during one revolution, varied its distance from the sun between 30 and 50 a.u., exceeding that of Neptune except at perihelion. He obtained greatly improved values for the longitude of the node and the inclination, but he noted that the values for the eccentricity and the longitude of perihelion would have to be corrected slowly by future observations.

Banachiewicz (1930a) noted that the determinations of Pluto's orbit which were based on the most widely-separated observations of 1930 were those of Lowell and Paris Observatories. Both of these indicated a very eccentric ellipse and a motion of recession from the earth. But the positions of 1919 to 1927 indicated an asteroidal eccentricity and a motion of approach. Banachiewicz found that if a different method of calculation was used, the observations of 1930 confirmed the earlier observations and he concluded that the orbit of Pluto resembled that of an asteroid and that the planet was approaching us. Banachiewicz (1930b) calculated the orbit of Pluto from the observations of January 23, February 21, March 21, and May 17, 1930 and found that his elements agreed well with those obtained by Miller (1930) and Stoyko (1930b) except for the eccentricity and the semi-major axis. Banachiewicz noted that his calculations and those of Bower and Whipple and

Nicholson and Mayall support the conclusion that Pluto is the trans-Neptunian planet sought by Lowell, or at least the first of several, as predicted by Lau.

Slipher (1931) reported that observers at Lowell Observatory with the 60-cm refractor, from observations of a distant terrestrial target of varying diameter and illumination, concluded that because of Pluto's faintness, its disk could escape detection in good seeing even if as large as 0".6. Their experiment was repeated, with similar results, by Alter (1952). Bower (1931) reported that workers at Lick Observatory, with the 90-cm refractor, found that the maximum diameter of Pluto, even allowing for some limb darkening, could hardly exceed 0".3. Nicholson at Mt. Wilson, with the 250-cm reflector, was unable to detect a disk in poor conditions and was therefore only able to report that it did not exceed 0".4. However because the effect of darkening at the limb was reduced by the large light-gathering power of the 250-cm instrument, it was concluded that the diameter could scarcely exceed 0".1.

Bower (1931), taking 0".43 as the earth's apparent diameter at Pluto's distance and assuming only that Pluto's density was equal to the earth's, calculated Pluto's mass to be 0.34 earth masses using the Lick estimate of a 0".3 disk for Pluto or 0.01 earth masses if the Mt. Wilson estimate of 0".1 was correct. Bower concluded that Pluto's mass could not be satisfactorily determined gravitationally at that time. From observations of magnitude and the most favorable assumptions of albedo and density, he placed an upper limit for the mass of Pluto at 0.7 earth masses. However, he noted that, until a disk was actually seen, the most probable value of Pluto's mass was 0.1 earth masses.

Possibly presaging his later change in attitude (see Chapter 4) Pickering (1930b) published a "supplementary note" in which he had a few words to say regarding Lowell's investigation of the new planet. Pickering noted that although Lowell's publication was "undoubtedly a very elegant mathematical investigation, showing high mathematical ability and great resource," it seemed to him that the mathematics were out of place and should be replaced by common sense. Pickering continued:

> Lowell and I were at one time more or less personal friends, and throughout his life I backed him in his Martian work far more fully, I believe, than did any other professional astronomer, because I believed that

he was largely right. It was only because I was unwilling to back him in all his theories, and because I said so, that we later, to my regret, became more or less estranged. He was glad and appreciative of my favourable comments on his work, but hurt by those of my criticisms that were unfavourable.[5]

---

[5] Pickering, W.H., The trans-Neptunian planet. A supplementary note. *Pop. Astron.*, 38, 293 (1930b).

Chapter 4
Filling in the Blanks, I
1931-1956

Bower and Whipple (1930a) published six sets of preliminary elements based upon the few observations of the Lowell Observatory Object (Pluto) then available. They found that a distinctly trans-Neptunian orbit satisfied the positions received at the time, but they noted that the observations could also be satisfied by parabolas. They estimated the object's diameter as being 1.4 times that of the earth from an assumption of a visual magnitude of 15, a distance of 41 a.u., and an albedo of 0.07. They noted that Mercury was the only other planet to have as low an albedo and that a larger albedo would yield a larger diameter so that it seemed probable that the object was smaller than the earth rather than larger. Unless the object's density was unusually great, the mass had to be much less than seven times that of the earth, Lowell's predicted value. Thus, as early as April, 1930 astronomers were aware of the contradiction between the apparent size of Pluto and that required by prediction to account for the observed perturbations of the other planets.

Baldet (1930) measured Pluto's diameter with the 80-cm Meudon refractor and reported that at most it could not exceed 6400 kilometers. Pickering (1931a) estimated Pluto's mass as being 0.71 earth masses and noted that this mass combined with Baldet's diameter would imply a mean density of 31 g/cm^3, or about one and a half times that of platinum. Furthermore, if Pluto's diameter was actually smaller, say 4800 kilometers, the mean density would reach 73 g/cm^3. Pickering commented that if the latter value was correct then "Pluto is the first planetary body discovered whose internal structure indicates that it had an origin outside of the solar system. Its density would thus seem to imply a relationship to those stars described as being of the white dwarf type."

Bower and Whipple (1931) published a second paper on the elements and ephemeris of Pluto based upon the numerous observations published since its discovery. Of special importance was the identification of Pluto on several pre-discovery plates including those taken at Uccle Observatory in January, 1927; Mt. Wilson, December, 1919; and Williams Bay, January, 1927 and January 1921. Bower's (1931) final set of elements, number XIX (Table 9), was based upon still more pre-discovery and post-discovery observations and was accepted as authoritative until revised by Cohen, Hubbard, and Oesterwinter (1967).

TABLE 9. Barycentric Elements XIX, 1900.0[1]

Epoch of osculation   1930 Sept. 20   G.C.T. = 242 6240 J.C.T.

T	1989 Oct. 0.0344	P	248.4301 5 tropic yr.
ω	113° 31' 17".72	a	39.5177 38
Ω	108 56 16 .18	p	37.0746 04
i	17 8 48 .40	μ	$\overset{\circ}{.}$0039 6750 28
q	29.6918 98	$k\sqrt{m_0}$	.0172 0215 04
e	.2486 438	$k\sqrt{m_0}$	$\overset{\circ}{.}$9856 1062

$$x = -27.6234\ 87\ (\cos E-e) + 25.2054\ 08\ \sin E$$
$$y = -28.2527\ 84\ (\cos E-e) - 24.3091\ 08\ \sin E$$
$$z = -\ 0.6121\ 78\ (\cos E-e) - 15.4550\ 70\ \sin E$$

---

[1] Bower, E.C., On the orbit and mass of Pluto with an ephemeris for 1931-1932. *Lick Obs. Bull.*, 15, 171, 1931. By permission of *Lick Observatory Bulletin*.

In his 1930 paper, E.W. Brown had commented that it was difficult to understand why predictions of an exterior planet by Lowell, Gaillot, Lau, and Pickering were possible from the very small residuals which the longitude of Uranus exhibited. He noted that the mass of the newly discovered Pluto was insufficient for the planet to have been predicted from its effect on Uranus and therefore its discovery near the predicted place was purely accidental. Brown (1931) later amplified his earlier scepticism and demonstrated how the use of a formula:

$$\sin(x + \Delta x) + \sin(x - \Delta x) - 2\sin x = -(4\sin^2 \tfrac{1}{2}\Delta x)\sin x, \qquad (1)$$

or in its more compact form,

$$\Delta^2 \sin x + (4\sin^2 \tfrac{1}{2}\Delta x)\sin x = 0, \qquad (2)$$

could effectively eliminate from a continuous set of observations a periodic term with known period but unknown amplitude and phase. Brown found that when the same transformation was applied to the perturbations in longitude of a planet produced by the attraction of an exterior planet, it gave the curve obtained by plotting these perturbations against the time a characteristic form which remained substantially the same over a considerable range of the ratio of the distances of the planets from the sun. He also found that the transformed observations would sometimes show whether a given set of residuals could be accounted for by the existence of an unknown planet or not. Applying his formula to the case of Pluto, Brown once again determined that the residuals of the longitude of Uranus were of no value for the prediction of an unknown planet and therefore the discovery of Pluto near Lowell's predicted place was accidental. He obtained a new upper limit, of about one-half that of the earth, for the mass of Pluto. Brown then showed that the existing observations of Neptune were insufficient to predict the existence of Pluto whatever its mass might be and that at least another century of observations of Neptune would be needed to obtain a value for the mass of Pluto with a probable error of less than a quarter of the earth's mass.

Brown's criticisms touched off a controversy with those workers who felt that the predictions of Lowell and Pickering were too close to the mark to be merely coincidental. The case for both schools of thought was examined in depth by Kourganoff (1941) as related by Reaves (see page 57).

Whereas Pickering (1931a) had cited Brown's (1930) earlier
paper to support his criticism of Lowell's prediction, he now
(1932c) took exception to certain parts of Brown's (1931) new
paper. Pickering thought that Brown confused the "two quite
distinct phenomena" that occur when one planet is approached
by another; one being the "maximum deviation", the other the
"maximum observable effect". Pickering noted that when two
planets pass one another the perturbing force or acceleration
is much greater than when they are nearly on opposite sides of
the sun, but it has less time to act so that the maximum deviation may be less. He then explained the maximum observable
effect as being the sharp cusp in the curve of plotted residuals when the two planets are in conjunction whereas in the
case of a perturbation that occurs near opposition, while the
deviation produced may be larger and the duration longer, *no
sudden change of speed occurs*. Indeed, such a gradual perturbation if plotted as a curve hardly shows at all. Pickering
concluded his rebuttal of Brown by stating: "it is very regrettable that his statements based on his theory do not seem
to coincide with the facts...."

It is interesting to digress here a bit and observe the
curious change in Pickering's attitude towards Lowell. In
one paper, Pickering (1931e) noted that although neither he
nor Lowell ever saw any of the plates containing images of
Pluto and, therefore, were not responsible for finding the
planet upon them, Lowell had at least left money to have the
search made. Because he had not made any similar gift, Pickering concluded:

> If it was a mere accident, as some claim, that his
> (Lowell's) position was so near to the actual one, then
> it was certainly a very fortunate accident for the finding of the planet. Had it not been for his bequest, it
> is quite probable that the planet would not have been
> found for many years, and he should therefore receive
> full credit for this very important part in the discovery.[2]

---

[2] Pickering, W.H., The discovery of Pluto. *Mon. Not. R.
Astron. Soc.*, 91, 812 (1931). By permission of the editor
of *Monthly Notices*.

Perhaps piqued by the "surprising and reckless claims to the exclusive prediction of its location put forth by his (Lowell's) active adherents and administrators, backed up by very extensive newspaper propaganda," Pickering (1931a) commented:

> No astronomer that I can recall has pointed with pride to Lowell's success in locating the planet at a longitude differing by a little over 5° from its actual place in the sky... No one who has carefully studied his work has expressed confidence that even this very modest success was due to other than mere accident.[3]

Bower (1932) observed that Pluto's mass was too small to be found gravitationally with any degree of confidence from the residuals yet available of any planet. He suggested that Pluto's mass might be about equal to Triton's, *i.e.* 0.06 - 0.09 earth masses, since if the latter was removed from Neptune's distance out to Pluto's distance of 40 a.u., its visual magnitude would increase from 13.5 to 14.2 which was just that found for Pluto. Crommelin called attention to an error in this magnitude change: removing Triton to Pluto's distance would change its magnitude to 14.9. Hence Bower (1933) gave a revised estimate of Pluto's mass of 0.1 to 0.3 earth masses. Bower (1934) used Baade's (1934) determination of Pluto's photographic magnitude of 14.9 as confirmation of the estimate of Pluto's mass being the same as Triton's.

Jackson (1930) recounted how Neptune by 1904 had fallen about 5" behind the place predicted by Leverrier's theory and since 1904 the planet had fallen about 3" behind Newcomb's (1899) theory which had replaced Leverrier's. Jackson showed that by a simple change of the elements, by adding three hours to the period, not only the modern observations, but also Lalande's 1795 observations could be satisfied. Pickering (1933) pointed out that the correction is, of course, due to the forward pull of Pluto on Neptune, increasing the speed of Neptune and thus apparently shortening its period. He provided tables giving Newcomb's and his deviations from Neptune's computed positions for the interval 1795-1896 and Jackson's and his deviations for the interval 1898-1929.

---

[3]Pickering, W.H., The mass and density of Pluto. *Pop. Astron.*, 39, 2 (1931).

Pickering believed that where the irregular variation of the deviations was as large in proportion to the deviations themselves as in the tabulated data, more satisfactory results could be obtained by his graphical method than by analytical methods. Pickering suggested that Jackson might have predicted the location of Pluto from his curve as accurately as he had from his own curve. If Jackson had completed his paper two or three years earlier and had examined his photographic plates with sufficient care, he might have "secured a very large fraction of the credit for the discovery of the ninth planet for England."

Nicholson and Mayall (1931) computed new orbital elements for Pluto using positions of Pluto obtained from photographs taken by M.L. Humason in 1919 with the 25-cm Cooke triplet and from photographs taken with the 150-cm and 250-cm Mt. Wilson reflectors in 1930. They also used Jackson's (1930) summary of the differences between the observed longitudes of Neptune and those computed from Newcomb's elements to determine the mass of Pluto by a least-squares solution. They derived a value of 0.94 ± 0.25 earth masses for the mass of Pluto. Russell (1931) felt that the mass of Pluto determined by Nicholson and Mayall was sufficient to account for the perturbations of Uranus and Neptune and rather over optimistically concluded that "all contradictions vanish."

Pickering (1931b) described the orbit of comet Wilk, 1930 III, as interesting because it belonged to a group of some 15 or 20 comets whose aphelia were supposedly associated with his trans-Plutonian planet P. Pickering (1931d) believed that the location of planet P was about as accurately known as was that of Pluto before its discovery. He reported that its position for epoch 1930.0 was $\alpha = 20^h 08^m$, $\delta = -53°.9$. He predicted planet P would exhibit a disk over 1" in diameter and of 11th magnitude. He urged that a visual search be undertaken for a few nights in the most plausible region since this might possibly save a lengthy photographic search. Figure 4.1 compares the orbits of Pluto and planet P.

Pickering (1932b) reported having issued a circular letter to about 15 different observatories in the southern hemisphere, stating where he believed planet P to be located, its magnitude, and other pertinent information. Rather to his surprise he received five replies: two observers did not have suitable instruments for the work, one reported cloudy weather, and two failed to find the planet. J.S. Paraskevopoulos at the

Harvard South African Station, who had earlier obtained photographs of Pluto after its discovery, took plates showing stars fainter than magnitude 15 with the 25-cm Metcalf refractor. He reported that "unfortunately no strange object has been found." H. Horrocks of the Cape Observatory reported finding no evidence of an ultra-Neptunian planet down to a magnitude of at least 14.7 on plates taken with a 12.5-cm wide-angle lens camera.

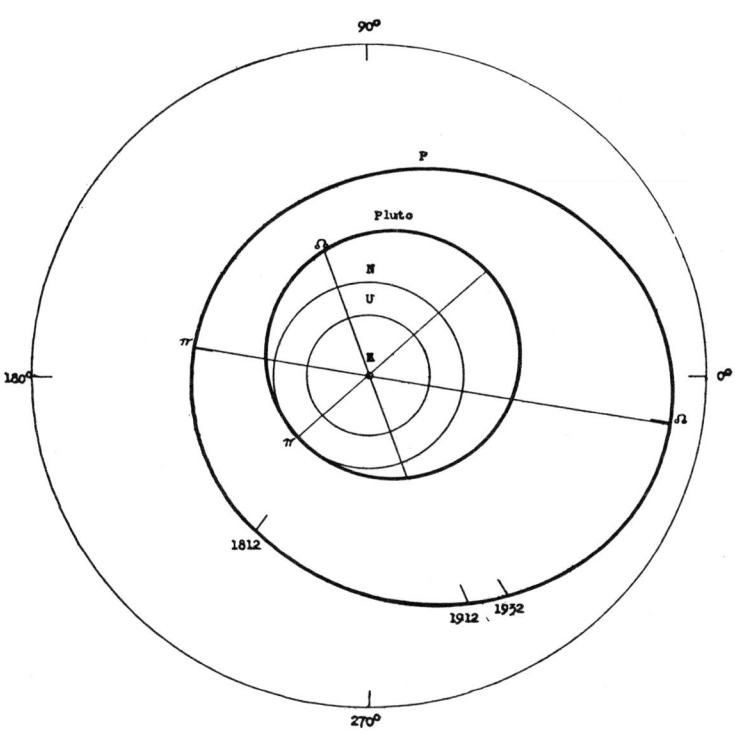

Fig. 4.1. The orbits of Pluto and planet P^4

---

[4]Pickering, W.H., Planet P, its orbit, position, and magnitude. Planets S and I. Pop. Astron., 39, 385 (1931c).

Pickering (1932a) also predicted a planet U with an orbit between the orbits of Jupiter and Saturn on the basis of the residuals of longitude of those planets. He (1931c) also postulated a planet S with a period of 328 years, a longitude of 347°, and an inclination of 30°.5; and a planet T with an orbit between those of Uranus and Neptune.

Fayet (1931a) examined the orbits of all comets appearing between the years 1680 and 1930 and found that only one comet out of some 240 had made a close approach to Pluto, this occurring in 1920.

In much the same vein as Forbes (1880a,b) and Flammarion (1884), C.H. Schuette (1949) noted the correlation between the aphelia of families of comets and the corresponding giant planets and suggested that there was evidence for the existence of a planet beyond Pluto which he called "Transpluto" at a mean distance of 77.2 a.u. Schuette drew a circle with a radius of 77.2 units and added the orbits of eight comets of the transpluto family drawn to scale (Fig. 4.2)

He pointed out that it was remarkable that the planes of the orbits of these eight comets rose above the ecliptic in the same direction, indicating a high inclination for the orbit of Transpluto. He doubted that it would be possible to determine the orbit of this planet from any perturbations of Uranus, Neptune, or Pluto. In addition it probably would be so faint that its discovery might only happen by chance.

Fayet (1931b) found a period of a little less than 500 years for the interval between successive heliocentric conjunctions of Pluto and Neptune employing the elements of Nicholson and Mayall (1931) and neglecting perturbations. Roure (1934) found that the mean annual motion of Pluto did not differ greatly from one-third of that of Uranus; the period of the inequality being 6516 years. Roure (1939) calculated a period of about 21,000 years for the secular inequality in Pluto's mean motion due to the perturbations of Jupiter and Saturn. He later (1940) found a period of about 12,000 years for the secular change in the eccentricity of Pluto's orbit brought about by the perturbing effect of Neptune.

Although Cohen and Hubbard (1964, 1965) later demonstrated the impossibility of such an event, Lyttleton (1936) considered that encounters of Pluto with the Neptunian system may eventually occur and may have occurred in the past. He examined

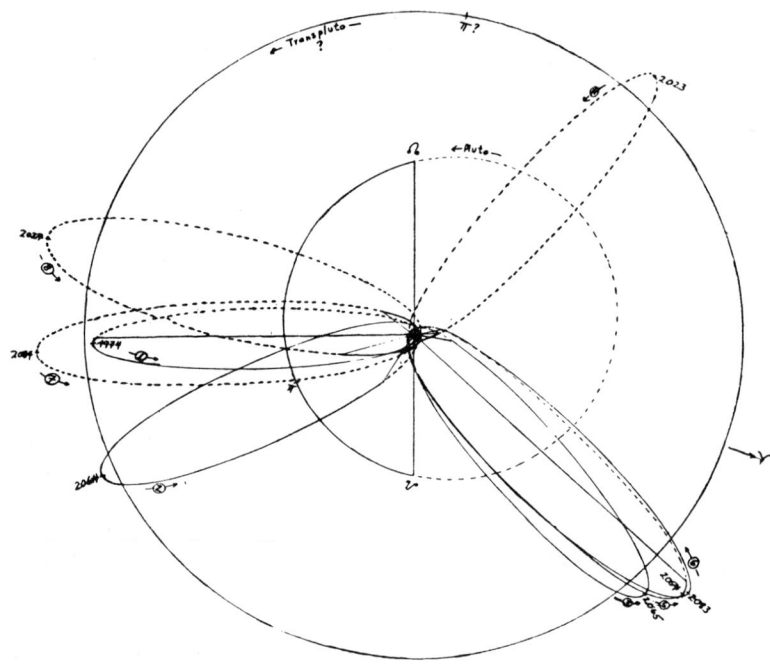

Fig. 4.2. Eight Comets of the Transpluto Family[5]

        1 = Comet 1851 IV (C.H.F. Peters)
        2 = Comet 1932 X (Dodwell-Forbes)
        3 = Comet 1937 III (Nagata)
        4 = Comet 1885 III (Brooks)
        5 = Comet 1905 III (Giacobini)
        6 = Comet 1932 I (Houghton-Ensor)
        7 = Comet 1932 V (Peltier-Whipple)
        8 = Comet 1874 IV (Coggia)

---

[5]Schuette, C.H., Two new families of comets. *Pop. Astron.*, 57, 176 (1949).

several different types of encounters, particularly that one where Triton and Pluto were both initially direct satellites of Neptune. Lyttleton commented that if Pluto had always been an independent planet, it would be natural to expect that its orbit would lie completely outside those of the others. On the other hand, supposing Pluto and Triton to both have been direct satellites of Neptune, a mechanism had to be found to bring them close enough together for an encounter to occur. Lyttleton suggested that tidal friction could have caused the satellites' mean distances to approach equality, resulting in an encounter, and he noted, in connection with this, that the rotation period of Neptune was the longest of the major planets. He considered the relative energies involved in such an encounter and concluded that Pluto may originally have been a direct satellite of Neptune, and that the encounter which gave it independence as a planet also reversed the general direction of Triton's motion.

Dauvillier (1951) suggested that Pluto and Triton are actually twin members of the family of giant planets which have been reduced to a dense core like the terrestrial planets because of insufficient initial mass and a high initial temperature which resulted in the loss of their lighter elements.

Putnam and Slipher (1932) noted that the distinctly yellowish color of Pluto observed by C.O. Lampland with the 105-cm Lowell reflector indicated a different atmosphere from that which envelopes Neptune and that, Pluto, like Mercury, may be without an atmosphere. They also pointed out that owing to the considerable eccentricity of Pluto's orbit the planet's brightness will vary considerably. They said that Pluto was then slowly increasing in brightness and would continue to do so until the year 1989, after which time Pluto would begin to dim for the following 125 years. This increase in brightness was confirmed by the visual estimates of Pluto's magnitude made by Moseley (1969), but contradicted by the photometric observations of Hardie (1965a, b) which indicated a dimming of the planet.

After the discovery of Pluto it was decided (Slipher, 1938) at the Lowell Observatory that the planet search should be continued to extend around the ecliptic and to higher and lower declinations. The plates taken in 1930 had covered a single strip centered along the ecliptic. After the search was expanded, two strips parallel to the ecliptic, one north and one south of it, were photographed. By September, 1932 a complete belt 30°-35° wide, straddling the ecliptic, had been both photographed and blink-examined. By May, 1943 the entire sky visible from Flagstaff, from 50° south to the North Celestial Pole, had been photographed with three or more plates of each region and an exposure time of one hour (Tombaugh, 1961). The plates taken with the 12.5-cm Cogshall camera reached 14-15th magnitude while the 25 by 42.5-cm plates reached as faint as magnitude 18.6. From July, 1943 to August, 1945 search work was interrupted by war research, but resumed from August to November, 1945 with the blink examination of some plates in northern areas. Since then no further searching was done until the work of Foss, Shaw-Taylor, and Whitworth (1972). Figure 4.3 shows the area of sky searched by 1945.

Tombaugh expressed the hope that the remaining unexamined photographed areas could be blinked at a future time. Such a blink examination might prove fruitful; see Brady (1972).

A total of 45,117 square degrees, or 75.4m^2 of photographic plate, including overlap areas, were blinked, involving some 7000 hours of work. The estimated number of stars in the examined areas was 44,675,000, or an average of 1000 stars per square degree. Tombaugh (1961) expressed his belief that the planet search work was done with such care and thoroughness that no unknown distant planets brighter than the 16th magnitude exist and that any planet between magnitudes 16 and 17 had a good chance of having been discovered. However, he noted that it would be possible to extend the search to the 20th magnitude with the 120-cm Schmidt telescope at Mt. Palomar. This instrument would permit the detection of an object 600 kilometers in diameter with an assumed albedo 2½ times that of the moon. But he also noted that the number of stars that would be recorded down to magnitude 20 would be overwhelming in regions near the Milky Way so that any planetary search would have to be confined to non-Milky Way regions. He suggested setting up two patrol areas, in Aquarius and eastern Cancer, which would be photographed each year at opposition season to catch possible objects as they merged from the Milky Way.

Shechtman (1945) reviewed Einstein's prediction of the advance of the perihelion of Mercury. He noted that Mercury was subject to planetary perturbations and that their existence raised doubt as to the "sufficiency" of perturbative residuals as a decisive test of Einstein's prediction. He pointed out that Pluto, on the other hand, was located at the outer limits

Filling in the Blanks, I 55

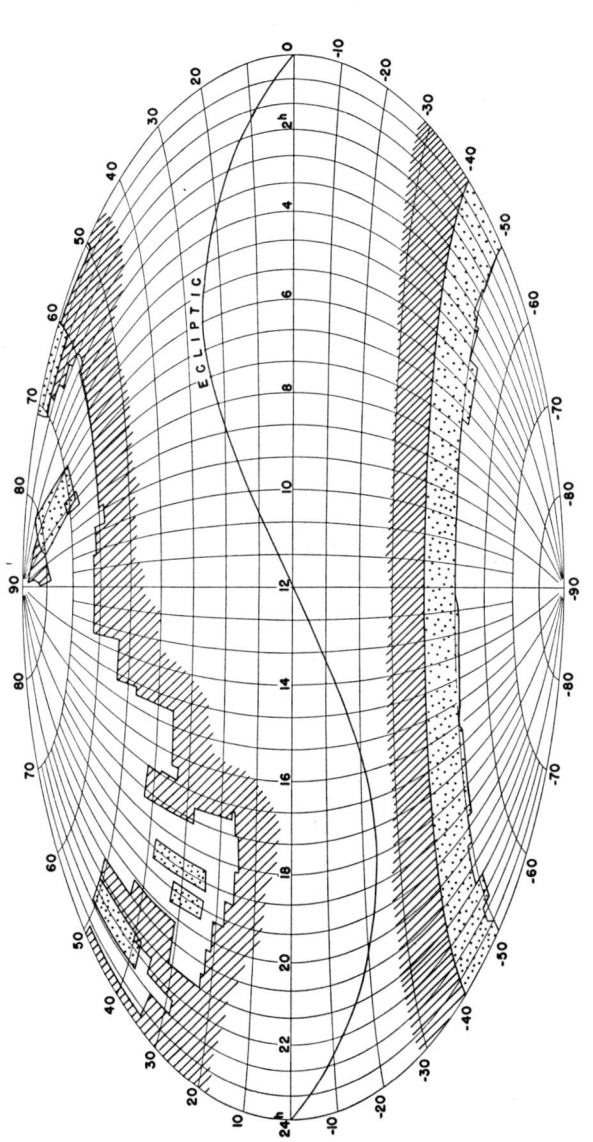

Fig. 4.3. Area covered by trans-Neptunian planet search, Lowell Observatory, 1929-1945.[6]

Dark shading: limiting magnitude 16-17.
Light shading (mostly between -40° and -50° declination): limiting magnitude 14-15.

---

[6]Tombaugh, C.W., In: *Planets and Satellites*, G.P. Kuiper and B.M. Middlehurst, Eds., p. 12. Univ. of Chicago Press, Chicago, 1961. With permission of author and publisher. Copyright 1963 by The University of Chicago. Published 1963.

of the solar system and was thus free of the perturbative complications of Mercury's apsidal motion. In addition, Pluto has an even larger eccentricity than Mercury so that Pluto's advance of perihelion should provide a more crucial test of Einstein's prediction. Shechtman concluded that Pluto's unique characteristics of a virtually unperturbed and uncomplicated motion of perihelion made an observational check upon its apsidal motion especially desirable.

In 1948 and 1949 Gerard Kuiper attempted to measure the diameter of Pluto with the 205-cm telescope at McDonald Observatory, but owing to the faintness of Pluto no reliable results were obtained in spite of considerable effort. Figure 4.4 shows two photographs of Pluto obtained by Kuiper with the 205-cm McDonald instrument. Kuiper (1950) later succeeded in measuring Pluto's diameter with the 500-cm Hale telescope on Mt. Palomar. In collaboration with M.L. Humason, Kuiper attached an illuminated disk-meter to the prime focus of the telescope and observed Pluto at an effective magnification of 1140 times. He found the diameter of Pluto to be 0.021 mm, or 0".23. He did comparisons with illuminated artificial disks and established that the preceding result was a real measure and not merely an upper limit.

Kuiper computed Pluto's albedo from Baade's (1934) value for the photographic magnitude to be 0.17, a much more reasonable value than the 0.04 value based on the assumption of Pluto being the same size as the earth. Kuiper calculated Pluto's diameter to be 0.46 times the earth's and felt that such a planet must have an atmosphere, though most of its original atmosphere would have frozen out. He noted that an atmosphere and a snow or ice-covered surface would prevent the albedo from being very low. However he noted that the albedo need not be that of clean snow, 0.7-0.8, because the snow would likely have been darkened over the ages by grit deposited by comets and meteors. He estimated Pluto's atmosphere to be equivalent to 0.1 terrestrial atmospheres.

Kuiper (1953) pointed out that if the rotation period of Pluto was found to be of the order of a day or less, it would be very probable that Pluto had formed as a planet; if the period was of the order of a week or so, it almost certainly formed as a satellite of Neptune, showing the slow rotation of an outer satellite. In any event, he deemed Pluto's mass to be too small to have perturbed many comets from the comet zone outside Neptune.

Walker and Hardie (1955) made a photometric determination of
the rotational period of Pluto, twenty-five years after
Russell (1930a) suggested such a determination. They used observations made by Walker in 1954 with the 150-cm and 250-cm
reflectors at Mt. Wilson and in 1955 by Walker and Hardie with
the 105-cm Lowell reflector. Kuiper contributed some 1953 observations made with the 205-cm McDonald Observatory instrument. The brightness of Pluto, in the 1954 and 1955 measurements, was compared with that of two near-by comparison stars,
selected to be about two magnitudes brighter than Pluto to
reduce the required observation time and of very nearly the
same color as Pluto so that neither the differential zenith
distance nor the differential color terms needed to be taken
into account. Walker and Hardie arrived at a value of 6.390
days ± 0.003 day for the rotational period. They suggested
that future observations might establish whether or not the
shape of the light curve was invariable and also whether the
amplitude was a function of the position of the planet in its
orbit, as this might indicate the inclination of Pluto's axis
of rotation. They suggested that Pluto is presently seen more
nearly equator-on than pole-on since its integrated light
changes by 0.1 magnitude as the planet rotates which is also
the light variation exhibited by Mars which is, of course,
seen equator-on.

As noted by Reaves (1951), most accounts of the discovery of
Pluto do not mention the fundamental contributions of Kourganoff (1941). Shortly after Pluto's discovery, Brown (1930,
1931) published two papers in which he concluded that the
calculations of Pickering and Lowell were irrelevant and that
the discovery of Pluto near Lowell's predicted place was
accidental. Perhaps because a detailed analysis of Lowell's
work would have been tedious and Pickering's method was empirical rather than analytical, Brown did not attempt to
analyze either's work. Instead he couched his indirect criticisms in four main arguments:

> 1) A prediction such as Lowell's, depending on old
> observations of Uranus (before 1780) because of the
> smaller residuals of modern observations, is of no
> value because of the large probable errors in the
> older observations.
> 2) The actual mass of Pluto turned out to be much
> less than Lowell's predicted mass and as a result
> Pluto's perturbations of Uranus are less than the
> probable errors of the pre-1780 observations.

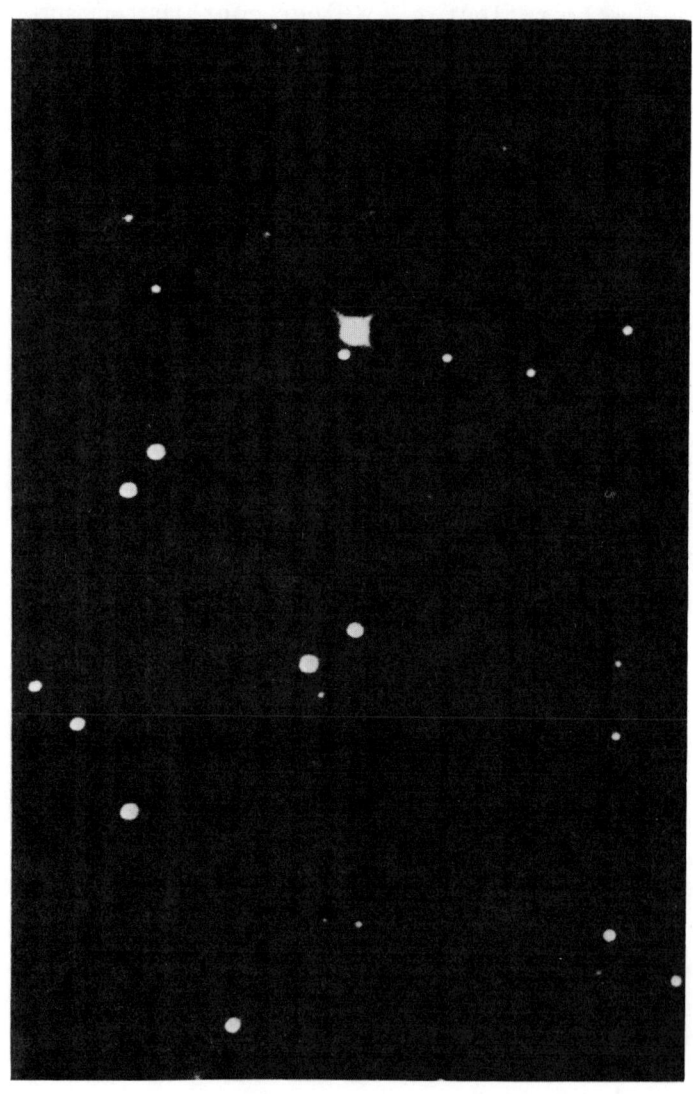

Filling in the Blanks, I 59

Fig. 4.4. RAS 577. G.P. Kuiper. McDonald Observatory. 2.1 m Refl., prime focus. *Above*, 1950 Jan. 24. 7^h 01^m U.T. Exp. 30 mins, full aperture. 103a-O plates. *Below*, 1950 Jan. 25. 6^h 50^m U.T. Exp. 60 mins, 1.4 m diaphragm. Note that Pluto has moved during the second exposure. Many faint galaxies are shown.[7]

---

[7]Courtesy of McDonald Observatory.

3) Lowell predicted a conjunction between Pluto and Uranus in 1853 simply because that date was near the middle of the modern observations.
4) The modern residuals of Uranus do not exhibit the general characteristics of having been caused by the perturbations of a trans-Neptunian planet.

Brown did not address himself to certain questions arising from his criticisms which other workers, Kourganoff in particular, asked. Namely, why was Lowell's solution determinate; why did the predictions of Lowell and Pickering agree so closely; and why did Lowell's planet X reduce the systematic residuals of Uranus by 99 percent? Kourganoff divided his study of the problem into five subject areas: I. Pluto's Perturbations of Uranus; II. The Work of Lowell; III. The Work of Pickering; IV. Discussion of the Criticisms of Brown; V. General Conclusion.

After considering the evidence for Pluto's perturbations of Uranus, Kourganoff concluded that the pre-1780 perturbations were significantly larger than the modern perturbations, but not entirely because of large systematic errors in the older observations. Kourganoff agreed with Brown that the modern perturbations of Uranus were too small for the prediction of any of Pluto's elements. To illustrate how close Lowell's prediction was, and by implication how improbable any "accident" or "coincidence" was, Kourganoff drew a diagram presented here in Fig. 4.5.

Lowell obtained two diametrically opposite solutions which reduced the residuals of Uranus by 99 percent and 90 percent respectively. Although Lowell felt this difference of 9 percent was insufficient to favor one position over the other, Pluto was actually found near the position giving the 99 percent reduction.

Pickering's method was a graphical one and hence of an empirical and crude nature; yet, as Kourganoff notes, he was able to make the following significant contributions:

1) In 1909 Pickering successfully predicted the next perturbations by the unknown planet from the residuals of Uranus.
2) In 1919 he predicted the position of a trans-Neptunian planet from the residuals of both Uranus and Neptune.

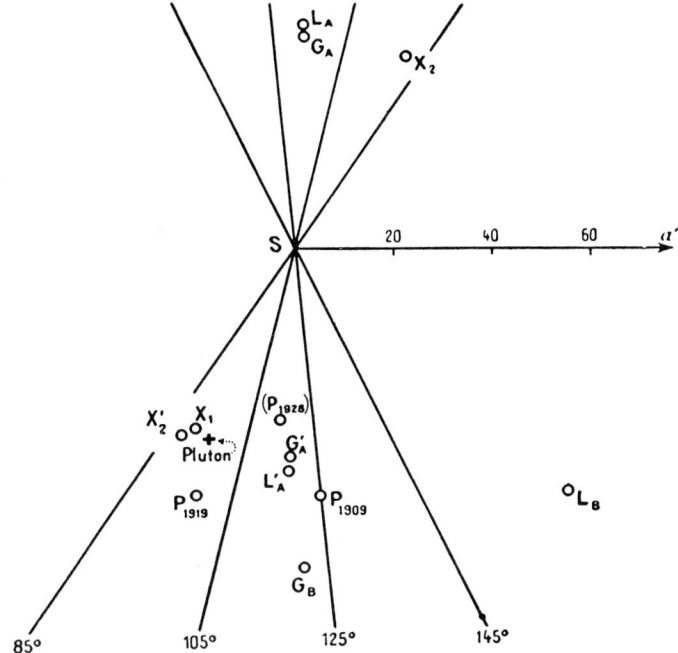

Fig. 4.5. Positions of Pluto and "theoretical" planets in 1920.0[8]

The positions are projected onto the plane of the ecliptic. The true position of Pluto in 1920.0 is marked by +.

Symbol	Predictor
X_1 and X_2	Lowell
L_A and L_B	Lau
G_A and G_B	Gaillot
P_{1909}, P_{1919}, P_{1928}	Pickering

The primes denote symmetrical positions. (This is Figure 21 in Kourganoff's memoir.)

---

[8] Reaves, G., Kourganoff's contributions to the history of the discovery of Pluto. *Publ. Astron. Soc. Pac.*, 63, 49 (1951). By permission of the author and the editor of *Publications*.

3) He removed the duality of Lowell's solution.
4) He successfully predicted the inclination and node of Pluto's orbit from Neptune's residuals of latitude.

In discussing Brown's criticisms, Kourganoff refutes his arguments individually. Firstly, he found that the pre-1780 observations, with a few exceptions, were much more accurate than Brown supposed. Secondly, Lowell's incorrect prediction of Pluto's mass did not affect his longitude prediction and while smaller than predicted, Pluto's mass was still sufficient for its perturbations to show in the pre-1780 residuals. Kourganoff considered in some detail the validity of Brown's third argument and concluded that Lowell's reasoning was sound. To simplify the comparison of the observed residuals of Uranus with those caused by a hypothetical trans-Neptunian planet, Brown devised a mathematical formula [Formula (1), see page 46]. Kourganoff questioned the validity of Brown's formula in his rebuttal of the latter's fourth argument. As Reaves (1951) observed:

> Kourganoff shows that Brown's use of the transformation as a *criterion* is unacceptable for several reasons.
>
> 1) Because of the character of the transformation, it is not directly applicable to isolated observations, that is, to the old residuals.
> 2) The assumption of circular orbits is made in deriving the transformation; this is a poor assumption for the orbit of Pluto which has an eccentricity of 0.25.
> 3) It is not sufficient to transform only the modern residuals because the perturbations of Uranus are sensible only before 1780 and will not begin to be significant again until after 1950-60.

Kourganoff modified Brown's transformation and used all of the observations. He found that for both Uranus and Neptune, the transformed residuals (in the neighbourhood of maximum perturbations) *do* exhibit a form characteristic of the perturbations of an exterior planet.[9]

---

[9]Reaves, G., Kourganoff's contributions to the history of the discovery of Pluto. *Publ. Astron. Soc. Pac.*, 63, 49 (1951). By permission of the author and the editor of *Publications*.

Regarding Kourganoff's general conclusion, we again quote Reaves:

> Kourganoff's conclusion is: The thesis of "pure chance" in regard to the discovery of Pluto is absolutely untenable. Pluto was "discovered" in 1915 by Lowell and rediscovered in 1919 by Pickering by the methods of celestial mechanics, before its "physical" discovery by Tombaugh at the Lowell Observatory.[10]

Alter (1952) noted that if Kuiper's value for the diameter of Pluto was correct and a reasonable value is assumed for the density, then the mass of Pluto is too small to have caused the perturbations of Uranus that led to its discovery. Alter put forth a possible explanation for this problem, suggested by Sir James Jeans and quoted by Crommelin (1934), that perhaps the surface of Pluto reflects light more or less as does a mirror. If so, we would observe sunlight reflected from only a small area near the centre of the surface and none from the limbs. A rough surface would allow us to measure the true diameter. Alter reported on an experiment in which four small balls of equal size, but with different kinds of surface, were photographed through a 30-cm refractor under darkfield illumination. The surfaces of the balls were, respectively: polished steel, matt-white, aluminum paint, and a combination of dark rough areas and patches of aluminum paint. The balls all appeared to be of different diameters and Alter concluded that in the case of a planet whose surface characteristics are unknown, no matter how carefully the diameter may be measured, we merely obtain a minimum value. The planet may be as small as the measurement, but, on the other hand, it may be far larger. The only apparent hope of making a true measurement is by the "extremely tedious" method of watching for it to occult stars. Even then quite a number of occultations would be necessary.

---

[10] Reaves, G., Kourganoff's contributions to the history of the discovery of Pluto. *Publ. Astron. Soc. Pac.*, 63, 49 (1951). By permission of the author and the editor of *Publications*.

Chapter 5
Filling in the Blanks, II
1957-1972

Kuiper (1956) noted that the assumption that Pluto formed out of the original solar nebula as a proto-planet was incompatible with its present orbital characteristics of high eccentricity and inclination. He ruled out the possibility that Pluto may have been perturbed afterward by a passing star by pointing out the low probability of such an event and that the nearly circular orbit of Neptune was apparently not perturbed. He instead proposed that Pluto originated as a satellite of proto-Neptune. Kuiper's hypothesis differs from that of Lyttleton's (1936) in which Pluto and Triton were initially both satellites of Neptune before having a close encounter which resulted in the retrograde motion of Triton and the ejection of Pluto from the system. Kuiper suggested that Pluto might have retained a long period of rotation from its Neptunian satellite origin. Photometric observations of Pluto begun in 1952 and culminating in the work of Walker and Hardie (1955) established a rotation period of 6.4 days.

Rabe (1954) developed the concept of the osculating restricted problem of three bodies and the osculating Jacobi integral,

$$V^2 = 2/r + r^2 + (2/p+p^2)\mu - C, \qquad (9)$$

where r and p are the satellite's distance from the sun and the primary, respectively, in units of the distance between the sun and the primary; C is the Jacobi constant; μ is the mass of the primary; and V is the satellite's velocity in the rotating coordinate frame of the restricted three-bodies problem.

Kuiper suggested to Rabe (1957a) the inclusion of the variable masses of proto-Jupiter and proto-Saturn in the solar mass M to facilitate the study of Pluto's orbital development under the combined mass decreases of Neptune, Jupiter, and Saturn. The inclusion of two variable masses, M and μ, means equation

(9) must be written as

$$v^2 = M(2/r+r^2) + \mu(2/p+p^2) - C, \quad (1)$$

where r and p denote the small body's distance from M and μ, respectively; V is its linear velocity in the rotating coordinate frame; and C is the variable Jacobi index.

To determine the combined effect of the mass decreases, dM and $d\mu$, on C, equation (1) becomes

$$dC = (2/r+r^2)dM + (2/p+p^2)d\mu. \quad (2)$$

Here dM and $d\mu$ are negative for decreasing masses and therefore C will decrease. If the motion of the small body is not limited to the plane of the two masses M and μ, another term for the z coordinate must be incorporated:

$$v^2 = M(2/r+r^2) + \mu(2/p+p^2) - (M+\mu)z^2 - C, \quad (3)$$

while equation (2) becomes:

$$dC = (2/r+r^2)dM + (2/p+p^2)d\mu - z^2(dM+d\mu). \quad (4)$$

Rabe (1957a) used equation (3) and values for r, z, and V provided by Eckert, Brouwer, and Clemence (1951) to compute C for the arbitrary date, July 10.0, 1921. The result, $C = 2.901$, and Pluto's present orbit are compatible with Kuiper's (1956) suggestion of a proto-Neptune origin, provided that the combined mass loss, dM, of the proto-planets was about 0.04 solar masses after Pluto escaped from Neptune. Kuiper (1957) estimated that the mass of Neptune itself had been reduced by a factor of 40 by solar evaporation before Pluto escaped.

Rabe (1957b) observed that while the above results were highly satisfactory, it did not follow that Pluto's initial mean motion as a planet was so close to the 2:3 commensurability with Neptune as today. He suggested that further orbital development after the escape from proto-Neptune resulted in the gradual attainment of the present near-commensurability. The proto-planet mass losses, dM from above, subsequent to Pluto's escape would account for the increase in the eccentricity angle, ϕ, from about 11° to the present 14.4°, as well as for the increase in the inclination, i, from an unknown initial value.

The proto-Neptune satellite hypothesis of Kuiper (1956), supported by the calculations of Rabe (1957a,b) would have occurred at an earlier stage of solar system evolution than the Triton-Pluto interaction hypothesis of Lyttleton (1936). Goldreich and Soter (1966) assumed that prior to their interaction both Pluto and Triton had direct orbits in or near Neptune's equatorial plane and that their separate tidal orbital evolutions could have eventually arranged a near collision. Provided that Pluto's mass is not too much greater than Triton's, such an encounter could have, in conserving momentum, reversed Triton's motion and ejected Pluto, imparting sizeable eccentricities and inclinations to both bodies. Goldreich and Soter point out that Lyttleton's hypothesis is supported by Walker and Hardie's (1955) determination of a 6.39 day period of rotation for Pluto since, if this was once Pluto's synchronous rotation period about Neptune, it would place Pluto at the time of ejection just outside the present orbit of Triton which has a period of 5.88 days. Tides raised on Neptune by Triton and particularly on Triton by Neptune would afterwards damp out any reasonable eccentricity imparted to Triton by the encounter in about 10^{17} seconds (~3 X 10^9 years).

Goldreich and Soter note that such a tidal mechanism would also damp out the eccentricity of Triton if it was captured into a retrograde orbit. They also point out that this enhanced plausibility of the Lyttleton hypothesis in turn lends credence to a less anomalous value for the density of Pluto than the ~50g/cm^3 required to account for the perturbations of Uranus and Neptune by a planet of the diameter determined by Kuiper (1950). They suggest that the perturbations might be attributable to a trans-Neptunian comet belt postulated by Whipple (1964). If, as required by Lyttleton's hypothesis, the mass of Pluto is not much greater than Triton's, say not more than double, then the density of Pluto would be less than ~2.3g/cm^3, a value which is consistent with those found for other satellites of the major planets.

McCord (1966) investigated the dynamical evolution of the Neptunian planet-satellite system in some detail and confirmed that a tidal friction mechanism could have reduced even a highly eccentric Triton orbit to its present form within the lifetime of the solar system. In fact, he concludes that Triton will continue to approach Neptune rapidly and will probably encroach upon the Roche limit within ten to one hundred million years and be destroyed. McCord finds a capture origin for Triton reasonable, but notes that a Triton-Pluto near collision is also tenable.

Cohen and Hubbard (1964, 1965) explored the orbital characteristics of Pluto by performing numerical integrations of the equation of motion of the five outer planets over the period extending backward from the present to more than 120,000 years ago. The computations were performed on the Naval Ordnance Research Calculator (NORC) at the U.S. Naval Weapons Laboratory at the rate of approximately 1500 years per hour of computer time. The data of Eckert, Brouwer, and Clemence (1951) were used. The computer program was designed to yield the geometry of each relative minimum of the distance between Pluto and Neptune.

The first close approach, AA', was encountered on calendar date August 27, 1896 at which time Pluto was at a true anomaly of 210°, 6° behind Neptune in true longitude and 19 a.u. distant. Over one synodic period of about 500 years, when Pluto completes two revolutions to three of Neptune, two other relative minima, BB' and CC', of 25 and 52 a.u. respectively, were observed. These occurred when Pluto was near successive perihelia. In successive synodic periods the positions of Pluto and Neptune were found to librate with a period of approximately 19,670 years as is evident in a plot of either the distance at the close approach AA' or the true anomaly of Pluto at closest approach. The latter is shown in Fig. 5.1.

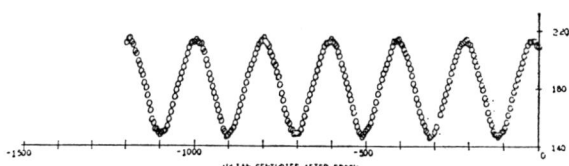

Fig. 5.1. True anomaly of Pluto at close approaches AA'.[1]

---

[1]Cohen, C.J. and Hubbard, E.C., Libration of the close approaches of Pluto to Neptune. *Astron. J.*, 70, 10 (1965). By permission of the authors and the editor of *Astronomical Journal*.

In Fig. 5.2, the positions and motion of the close approaches is summarized. The orbits in this figure are the projections on the invariable plane. The top row in the figure covers a synodic period when the libration of the close approach positions is at its first extremum, about 1500 years before epoch. This row shows the successive close approach positions of Pluto and Neptune, *AA'*, *BB'*, and *CC'*, in the synodic period. The second row shows the positions of *AA'*, *BB'*, and *CC'* during a synodic period 5000 years earlier and the third row is another 5000 years earlier. The three rows span half a libration period, before and after which the motion is reversed and then is periodically repeated.[2]

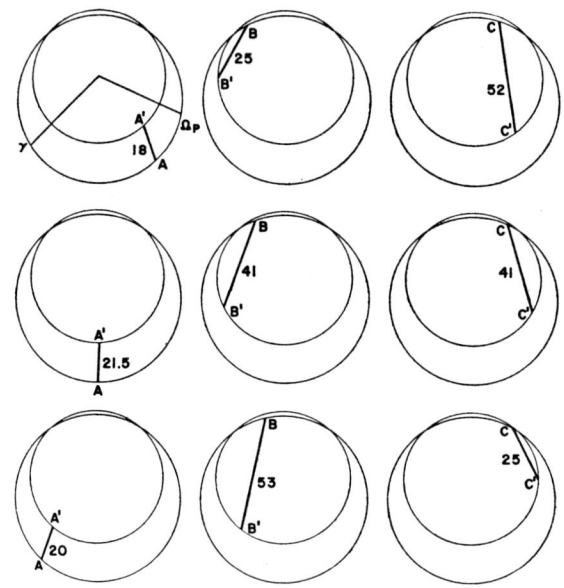

Fig. 5.2. Libration of close approaches; distances in a.u.[2]

---

[2]Cohen, C.J. and Hubbard, E.C., Libration of the close approaches of Pluto to Neptune. *Astron. J.*, 70, 10 (1965). By permission of the authors and the editor of *Astronomical Journal*.

The libration of the close approaches can be expressed in terms of classical variables by describing the deviation of the phases Pluto and Neptune from commensurability as the angle,

$$\delta = 3\lambda_P - 2\lambda_N - \tilde{\omega}_P - 180°,$$

where λ is the mean longitude, $\tilde{\omega}$ is the longitude of perihelion, and P and N stand for Pluto and Neptune, respectively. The period of the oscillation of δ was calculated to be 19,670 years, the mean about 180°, and the amplitude 76°.

Besides the mean anomaly, the other elements of Pluto exhibit osculation. The semi-major axis a shows a modulation with a period of 19,670 years and an amplitude of 0.14 a.u. The eccentricity e exhibits a decrease of approximately 0.005 in 120,000 years with a period of not less than a million years. The inclination i showed an increase of about 0.2° in 120,000 years, also with a period of not less than one million years. The node regresses at about 9°.5 per 100,000 years yielding a period of approximately four million years. The argument of perihelion at epoch is about 113°.6, increasing at a rate of about 1°.2 per 100,000 years with a period of 30 million years.

The libratory mechanism may be visualized with the aid of Fig. 5.3 which is explained by Cohen, Hubbard, and Oesterwinter (1979):

> The path of Pluto is plotted in a coordinate system with origin at the Sun and rotating with the angular speed of Neptune. The path of Pluto is traced for a synodic period, and the two loops correspond to two consecutive passages of Pluto through its perihelion. During a libration period of 80 revolutions of Pluto, the synodic path experiences a rotational oscillation about S with respect to the Sun-Neptune line. Instead of showing the libration of the path with N fixed, the figure shows the motion of Neptune N-N relative to the synodic path. Note that the direction of motion relative to the rotating frame, as shown in the figure, is opposite to the direction relative to the non-rotating inertial frame. Now consider the direction and size of Neptune's disturbing acceleration on Pluto by connecting Neptune's and Pluto's position when Neptune is at an extrema of its relative libration. Take the com-

ponent of this force which is perpendicular to the Sun
-Pluto line in order to consider the component which has
the principal effect on Pluto's motion. The time average
of this component around the synodic path is then easily
seen to be dominated by the component when Pluto is in
the perihelion loop nearer to Neptune. It is thus clear
that the mean component of tangential force accelerates
Pluto in the direction of its motion in the inertial
frame whenever Neptune is in the left half of the arc *N-N*
in Figure (5.3), and in a direction opposite to its
motion when Neptune is in the right half of this arc.
With the energy and period considerations made above, it
is clear that Neptune accelerates to the right when it is
in the left half of its arc relative to the synodic path,
and to the left when it is on the right. One may compare
this process to the motion of a pendulum inasmuch as Neptune oscillates back and forth on the arc *N-N*. The force
which causes this libratory motion of Neptune can be
thought of as a mutual repulsion between Neptune and
Pluto's perihelion loops.[3]

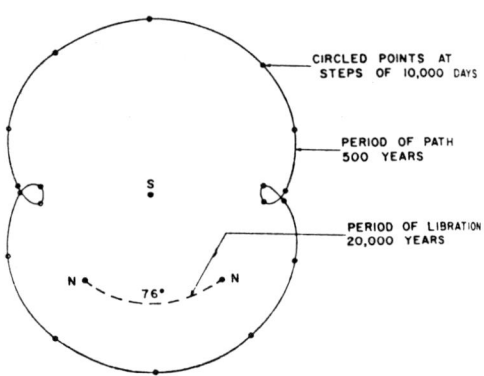

Fig. 5.3. Path of Pluto about *SN* at extrema of libration.[4]

---

[3]Cohen, C.J., Hubbard, E.C. and Oesterwinter, C., On the
stability of Pluto's orbit. (*Private communication*), 1979.
[4]Cohen, C.J. and Hubbard, E.C., Libration of the close
approaches of Pluto to Neptune. Astron. J., 70, 10 (1965).
By permission of the authors and the editor of *Astronomical
Journal*.

Because of the libration about the commensurability ratio, the closest approach Pluto can make to Neptune is about 18 a.u. and this is locked in near aphelion. Therefore an unusually close approach which would disturb the planets and which would have to occur near perihelion is ruled out. This also weakens the hypothesis that Pluto was once associated with the Neptune satellite-planet system.

Cohen and Hubbard (1965) used the orbit for Pluto published by Eckert, Brouwer, and Clemence (1951) which in turn was based on elements for Pluto published by Bower (1931). In 1967, Cohen, Hubbard, and Oesterwinter repeated the 1965 work, extending it to 300,000 years, and using an improved orbit for Pluto, but retaining the 1951 values for the other planets. The orbit of Pluto was revised by the method of least squares based on observations made between 1914 and 1965, some of which were previously unpublished. Fourteen pre-discovery observations were provided by Sharaf (1965) and 33 normal-places for the years 1930 to 1958 were obtained from Sharaf and Budnikova (1964). A total of 183 observations obtained with the 60-cm Yerkes and 215-cm McDonald reflectors were used. In addition, more than 300 observations made mainly by C.O. Lampland with the Lowell 105-cm reflector were used. Twenty-one Yerkes observations published by van Biesbroeck (1963) and nine positions used by Halliday and co-workers (1966) in their occultation studies completed the data sample. In the procedure employed to obtain weights for the various sets of observations, four separate solutions were made, one for each of the data groups, Sharaf-Budnikova, Yerkes-McDonald, Lowell, and Halliday.

Pluto's orbit and those of Jupiter, Saturn, Uranus, and Neptune were computed by numerical integration. The new elements obtained for Pluto, not surprisingly, differ significantly from those of Eckert, Brouwer, and Clemence (1951). Table 10 compares the two sets of elements.

Cohen, Hubbard, and Oesterwinter (1967) found the period of the oscillation to now be 19,440 years with an amplitude of about 80°. In 1971 they extended their integration to 1,000,000 years.

Williams and Benson (1971) integrated Pluto's orbit forward to 2.1 million A.D. and backward to 2.4 million B.C. They used the elements of Cohen and co-workers (1967) to start the integrations. For the integration of Pluto, they made several

TABLE 10. Old and New Elements for Pluto[5]

Element:	Cohen, Hubbard and Oesterwinter (1967) Epoch: JD 243.0000.5	Eckert, et al. (1951) Epoch: JD 243.0000.5	Cohen, Eckert, et al. Epoch: JD 243.0000.5
a	39.533 22 a.u.	39.518 11 a.u.	0.015 11 a.u.
e	0.246 0034	0.245 9386	0.000 0638
i	17°.122 886	17°.122 599	0°.000 287
M	289°.354 27	289°.279 52	0°.074 75
ω	113°.274 85	113°.342 53	-0°.067 68
$M+\omega$	42°.629 12	42°.622 05	0°.007 07
Ω	109°.606 471	109°.606 852	0°.000 381

[5]Adapted from: Cohen, C.J., Hubbard, E.C. and Oesterwinter, C., New orbit for Pluto and analysis of differential corrections. *Astron. J.*, 72, 973 (1967). By permission of the authors and the editor of *Astronomical Journal*.

simplifying approximations. The effects of the four terrestrial planets were approximated by including their mass in the mass of the sun. The value used for the mass of Pluto was 1,812,000 reciprocal masses (Duncombe, Klepczynski, and Seidelmann, 1968a,b). Table 11 gives the adopted masses of the outer planets.

TABLE 11. Inverse Masses of the Outer Planets[6]

Planet	1/mass($M_\odot$)
Jupiter	1047.35
Saturn	3501.6
Uranus	22869.0
Neptune	19314.0
Pluto	1812000.0

Williams and Benson confirmed and refined the libration period to 19,951 years over the 4.5 million year interval of their integration. They found the libration to be completely stable with the small variations in its mean value serving to actually increase the minimum distance of approach to Neptune and therefore the stability of the system. Assuming this stability over the length of the integration to be representative of the stability over the age of the solar system it appears that the Neptune-Pluto libration originated as early as the formation of the solar system. They concluded that an improvement in the model of the solar system would be desirable before any lengthier integrations of Pluto are undertaken.

The problems faced by Fixlmillner, Delambre, and Bouvard (see Chapter 1) in trying to calculate an orbit for Uranus that would satisfy both old and contemporary observations apparently remain just as intractable to modern workers attempting a solution to Neptune's orbit. Duncombe, Klepczynski, and Seidelmann (1968a,b) point to the apparent failure of past and current theories of the motion of Neptune to agree with obser-

---

[6]Williams, J.G. and Benson, G.S., Resonances in the Neptune-Pluto system. *Astron. J.*, 76, 167 (1971). By permission of the authors and the editor of *Astronomical Journal*.

vations other than those originally included in the computation of the theory. For example, Newcomb's theory (1899) which was in good agreement with the observations of 1795 and 1846-1896 failed to represent observations in orbital longitude by over 5 seconds of arc by 1938 even when amended to include perturbations by Pluto. The theory for Neptune's motion developed by Eckert, Brouwer, and Clemence (1951), incorporating a reciprocal mass of 360,000 for Pluto and fitted to the observations of 1795 and 1846-1938, departed from the observed longitude of Neptune in the 1960's by nearly 4 seconds of arc.

Duncombe and his colleagues (1968a,b) felt that these failures indicated a need for a different value for the mass of Pluto. They performed several numerical integrations of the orbits of the five outer planets over a range of assumed values for the reciprocal mass of Pluto: 360,000; 930,000; 1,500,000; and 2,640,000. The criterion adopted was that the best value for the mass of Pluto was that one which, when fitted to observations from 1846-1938, best represented the longitude observations for 1960-1968. In a move very reminiscent of Bouvard's (1821) rejection of ancient observations of Uranus (see Chapter 1), Duncombe and his co-workers omitted the 1795 observations from their calculations, basing their decision upon much the same reasons given by Bouvard.

Figure 5.4 shows the observed minus computed positions (0 - C's), or residuals, in orbital longitude, λ, and latitude, β, compared to the Eckert, Brouwer, and Clemence theory for a reciprocal mass of 360,000 for Pluto.

The same procedure was followed for the other assumed masses shown in Fig. 5.4. It was found that a reciprocal mass of 1,812,000 provided the best fit with observation. A final orbit obtained with the use of this mass of Pluto and fitted to the observations of 1846-1938 represents the observations of 1960-1968 (Fig. 5.5).

If Pluto's density is assumed to be the same as the earth's then the newly determined mass (= 0.18 earth masses) implies a diameter of 7200 kilometers. However, if the upper limit of 6400 kilometers for Pluto's diameter found by Halliday and co-workers (1966) is accepted then Pluto's density must be at least 1.4 times that of the earth.

Duncombe, Klepczynski, and Seidelmann concluded that any

Filling in the Blanks, II

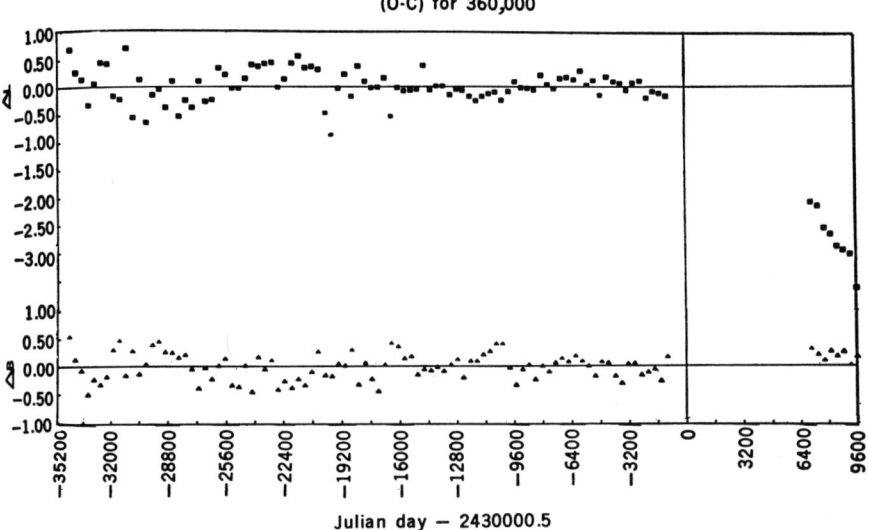

Fig. 5.4. The observed values minus the computed values (O-C's) in orbital longitude and latitude for the currently adopted reciprocal mass of Pluto.[7]

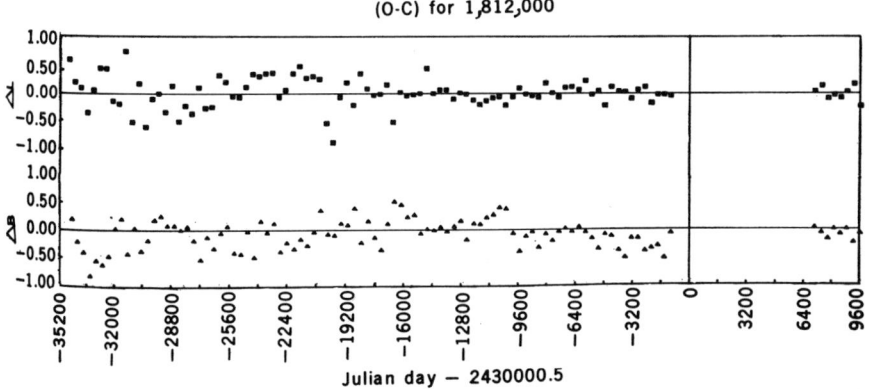

Fig. 5.5. Observed minus computed values (O-C's) in orbital longitude and latitude for the new reciprocal mass of Pluto.[8]

[7,8] Duncombe, R.L., Klepczynski, W.J., and Seidelmann, P.K., Mass of Pluto. *Science*, 162, 800 (1968). By permission of authors and editor. Copyright, 1968, by the A.A.A.S.

further refinement of the value for Pluto's mass and the elements of Neptune's orbit had to await the completion of observations of Neptune then being made at the U.S. Naval Observatory.

Halliday (1969) noted that the density for Pluto suggested by Duncombe and colleagues (1968a,b), namely 1.4 times that of the earth's, was more acceptable than the much larger values earlier suggested on the basis of higher assumed masses for Pluto. However, he observed that even this lower value was close to the density of iron meteorites, a composition for Pluto that he commented was incompatible with theories of the formation of the solar system. He suggested two reasons for questioning the mass values for Pluto derived from observations of Neptune:

(I) The longitude residuals may be influenced by undetermined systematic errors.

(II) The adopted values of the masses of Jupiter, Saturn, and Uranus may be too imprecise.

In connection with the second point (II) Halliday noted that even a small correction to the values for the masses of Saturn and Uranus might have sufficient effect on Neptune to allow for a still smaller mass for Pluto. Halliday concluded by suggesting a test calculation of Neptune's residuals using solutions where the masses of Saturn, Uranus, and Pluto were varied simultaneously.

Duncombe, Klepczynski, and Seidelmann (1970) repeated their 1968 analysis of Neptune's motion, incorporating the same observational data, but utilizing new reciprocal masses for Saturn and Uranus of 3498.7 and 22,692, respectively, published in a later paper (Duncombe, Klepczynski, and Seidelmann, 1971). A value for the reciprocal mass of Pluto of 1,900,000 (0.17 earth masses) was obtained which was not significantly different from the 1968 result (0.18 earth masses).

It will be informative to repeat the analysis of the motion of Neptune utilizing the much more refined values for the mass of Saturn obtainable from the celestial mechanics experiments of the Pioneer II and Voyager 1 and 2 space missions. Later it may be possible to refine the mass values for Uranus and Neptune if Voyager 2 is directed past those planets. However there are no space missions currently planned that would permit a direct determination of Pluto's mass.

Seidelmann, Klepczynski, and Duncombe (1971) followed up their earlier conclusion (Duncombe, Klepczynski, and Seidelmann, 1968a,b) that any improvement in the value for Pluto's mass would have to await completion of a discussion of observations of Neptune made at the U.S. Naval Observatory. They performed a simultaneous numerical integration of the orbits of the five outer planets, utilizing the newly revised values for the masses of Saturn and Uranus (Duncombe, Klepczynski, and Seidelmann, 1970) and a range of values for the mass of Pluto. The specific values for the reciprocal mass of Pluto used were 1,812,000; 2,500,000; 3,624,000; and 5,436,000. The best mass for Pluto was assumed to be the one that permitted the orbit of Neptune, when fitted to observations from 1846 to 1938, to best agree with the observations of longitude from 1939 to 1968. Disconcertingly, though not unexpectedly, they found a wide variation in the values of the reciprocal mass of Pluto determined from various spans of observations. They found that a reciprocal mass of 2,848,000 gave the best solution for the observations from 1939 to 1968 while if only the observations of 1960 to 1968 were used, the best value was 3,473,000. For this reason Seidelmann and co-workers chose to adopt the rounded value of 3,000,000 for the reciprocal mass of Pluto (= 0.11 earth masses) as being the most likely value based on the currently available observational data. This mass combined with a 6400 kilometers upper limit for the diameter of Pluto yields a density of 4.86 g/cm^3 or 0.88 times the earth's mean density. Assuming a density for Pluto the same as the earth's gives a diameter for Pluto of 6112 kilometers.

Ash, Shapiro, and Smith (1971) provided a list of the then most reliable estimates for the masses of all the major planets. They outlined the three principal methods of arriving at such estimates:

   (I) Celestial mechanics experiments by spacecraft flying past the planet

   (II) Application of Kepler's third law to observations of a satellite's period and mean distance from the planet.

   (III) Observation of resonance, long-term perturbation, and short-term perturbation effects of one planet upon another.

It can be noted that method (III) was, until recently, the only approach available for determining the mass of Pluto while method (II) became available in 1978 (Christy and Harrington) and method (I) will not be available for some considerable time yet. Ash and co-workers concluded that Pluto's mass could not be determined reliably from existing data by method (III) then the only approach available.

Kiladze (1968) gave a method for determining a planet's mass, taking into account the known radius and period of rotation by the formula:

$$\log [Q\sqrt{R}] = -0.93 \pm \frac{7}{3} - \log m - 0.17^{0.38} \qquad (1)$$

where Q is the planet's kinetic moment, R is the planet's radius in earth radii, and m represents the planet's mass with the earth's mass as unity.

If Q in formula (1) is expressed in terms of the mass, radius and period of rotation of the planet, and the mass given in terms of the density and radius, and the already known values of the orbital radius and the rotation period are substituted in the formula, then by discarding the minor term, the following relationship is obtained:

$$\varrho^2 r^3 \approx 0.05 \qquad (2)$$

where ϱ is the planet's density in units of the earth's density.

Kiladze noted that since Pluto's density probably does not exceed that of the earth, formula (2) yields a lower limit for Pluto's radius of 0.37 while the lower limit for the density was 0.6 times that of the earth's. A combination of these values gives a mass for Pluto of 0.05 to 0.09 earth masses. Kiladze considered a radius of 0.46 as being the most likely value which corresponds to a density of 0.72 (= $4g/cm^3$). The resulting mass of Pluto is then 0.07 earth masses. Kiladze also noted that if Pluto's period of rotation is not 6.39 days (Hardie, 1965), but twice as long, then the mass computed above will be reduced to 0.04 earth masses. He concludes that such a small mass for Pluto means that perturbations in the motion of Uranus and Pluto can be explained only by the influence of some unknown body, perhaps the tenth planet of the solar system.

Cameron (1962) suggested that as a result of the condensation of the primeval cloud of dust and gas into the Sun and planetary system there ought to be a large mass of small solid material on the periphery of the solar system. Whipple (1964) suggested that such material might exist in the form of a cometary belt. He thought that such a belt of comets while probably too dim to be visible from ground-based instruments might be detected by its gravitational influence on the outer planets. Whipple calculated that a belt of comets comprising perhaps 10 to 20 earth masses might exist in the region of 40 to 50 a.u. from the Sun.

In the case of Neptune the assumption of a cometary belt results in a slightly better fit with the latitude observations than does the assumption of perturbations by Pluto. Either assumption reduces the sum of the square of the residuals by more than a factor of three, a rather significant amount. For Uranus neither assumption results in a significantly better fit and neither assumption improves the fit in longitude very much for either Uranus or Neptune. Whipple points out that the mass of Pluto is too small to account for the perturbations of Uranus and Neptune and that this provides some support for the existence of a comet belt beyond Neptune.

Seidelmann (1971) investigated the possibility that the observational residuals of Uranus and Neptune could be explained by an unknown planet, or planets, beyond Pluto. He noted that such a planet had been suggested earlier by Schuette (1949), Naef (1955), Rawlins (1970), and Gunn (1970). He utilized the numerical integration program employed by Duncombe and co-workers (1968a,b) and others since it was easy to introduce additional planets into the calculations. Seidelmann selected three hypothetical planets, $S1$, $P1$, and $P2$, the first two being essentially the planets S and P proposed by Pickering (1931c); $P2$ is Pickering's planet P reduced to a more reasonable mass. Too large a mass for a hypothetical planet would imply a planet too large and bright to have escaped detection during Tombaugh's (1961) systematic searches. Tombaugh concluded that there is no undiscovered planet brighter than 16th magnitude and that any planet between magnitudes 16 and 17 had a good chance of being detected. The elements for Seidelmann's three hypothetical planets are given in Table 12.

The three hypothetical planets were separately introduced into

TABLE 12. Elements for hypothetical planets for the equator and equinox of 1950.0 and epoch 2430000.5[9]

	P1	P2	S1
Mean anomaly	141°26'51".898	141°26'51".898	141°26'51".898
Argument of perihelion	170°30'	170°30'	0°
Longitude of node	351°	351°	271°
Inclination	37°	37°	30°36'
Eccentricity	0.265	0.265	0.0
Semi-major axis (a.u.)	75.5	75.5	48.3
Reciprocal mass	7000	300,000	3,000,000
Magnitude estimate	13.6	16.1	15.7

[9]Seidelmann, P.K., A dynamical search for a trans-Plutonian planet. *Aston. J.*, 76, 740 (1971). By permission of the author and the editor of *Astronomical Journal*.

the calculations with the elements, masses, and magnitudes as listed in Table 12. The results of the integrations indicated that the hypothetical planet P1 would have so perturbed the outer planets that the residuals would be larger rather than reduced. In addition such a bright planet (magnitude 13.6) would almost certainly have been discovered by now. When the mass of P1 was reduced so that its magnitude would also be lowered to a more plausible level, it no longer exerted any noticeable effect on the residuals. Nor did S1, with a magnitude of 15.7, significantly affect the motions of any of the outer planets during the period of observation. Seidelmann concluded that while there may be some evidence of trans-Plutonian planets, particularly from the aphelia of some families of comets, such a planet would have little effect on the motion of the known planets except in the case of a close approach which is considered unlikely.

Brady (1972) contended that there is no logical reason to suppose that Pluto is the outermost planet in the solar system. He suggested that because the effect of a disturbing body is directly proportional to the eccentricity of the disturbed body, a trans-Plutonian planet would perturb a highly eccentric object such as Halley's comet much more noticeably than the major planets. Therefore he performed a numerical-graphical experiment to determine if it was possible to improve the orbit of Halley's comet by including the effect of a hypothetical planet beyond Pluto. Brady found that a hypothetical planet with the elements given in Table 13 would reduce the residuals in the time of perihelion for Halley's comet by 93 percent of Orbit 12a (1910-1456).

The effect of the hypothetical planet on the residuals of Halley's comet is shown in Table 14.

If the effect of a hypothetical planet on a comet is to be noticeable, the comet must have been observed at three or more apparitions and it must have a great enough aphelion distance for the effect to be detectable above the observational errors. Only two comets, Comet Olbers and Comet Pons-Brooks, meet these conditions. The orbits of these two comets were integrated similarly to Halley's comet, with and without the hypothetical planet. The effect of the planet is clearly discernable in the reduction of the residuals in Table 15.

The planet postulated by Brady would orbit the sun at twice Neptune's distance with an inclination of 120° and have a mass

TABLE 13. Elements for Brady's hypothetical trans-Plutonian planet for the equator and equinox of 1959.0 and epoch 2419326.5 (= 1911 October 16.5)[10]

Argument of perihelion	181°
Longitude of node	115°.75
Inclination	120°
Eccentricity	0.07
Semi-major axis (a.u.)	59.93575
Reciprocal mass	90,000
Period	464 years
Radius vector	63.491

TABLE 14. Residuals of Halley's comet with and without the trans-Plutonian planet[11]

Year of apparition	ΔT Orbit 12a 9 planets	ΔT Orbit 52B 10 planets
1910	$0^d.0$	$0^d.0$
1835	0 .0	-0 .8
1759	+4 .2	-0 .8
1682	+8 .2	-3 .0
1607	+42 .6	-2 .7
1531	+53 .5	+2 .5
1456	$+49^d.6$	$+3^d.9$

[10,11] Brady, J.L., The effect of a trans-Plutonian planet on Halley's comet. *Publ. Astron. Soc. Pac.*, 84, 314 (1972). By permission of the author and *Publications*; University of California, Lawrence Livermore Laboratory, and U.S. Department of Energy.

TABLE 15. Residuals of Comets Olbers and Pons-Brooks with and without the trans-Plutonian planet[12]

Year of apparition	ΔT Orbit A 9 planets	ΔT Orbit B 10 planets
Comet Olbers		
1815	$0^d.0$	$0^d.0$
1887	-0 .5	0 .0
1956	+5 .0	+1 .5
Comet Pons-Brooks		
	Orbit C	Orbit D
1954	0 .0	0 .0
1884	-4 .6	0 .0
1812	$-7^d.2$	$-4^d.1$

[12] Brady, J.L., The effect of a trans-Plutonian planet on Halley's comet. *Publ. Astron. Soc. Pac.*, 84, 314 (1972). By permission of the author and *Publications*; University of California, Lawrence Livermore Laboratory, and U.S. Department of Energy.

three times that of Saturn and a probable brightness of 13th or 14th magnitude. How could such a planet have eluded detection by Tombaugh's search? Brady points out that during the 16-year period of the search, from 1929 to 1946, the entire path of the planet lay outside the area covered by the search (Fig. 5.6).

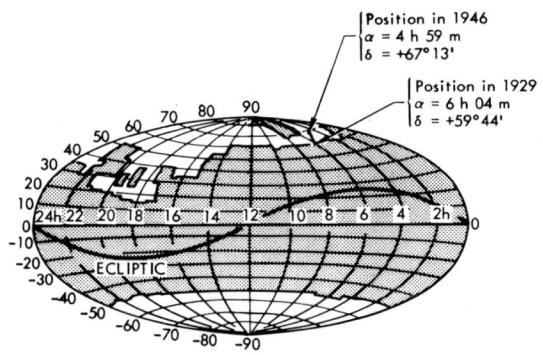

Fig. 5.6. Area covered by trans-Neptunian planet search made at the Lowell Observatory, 1929-45 (shown stippled). The positions of the hypothetical trans-Plutonian planet are shown for 1929 and 1946.[13]

Foss, Shawe-Taylor, and Whitworth (1972) conducted a photographic search for Brady's (1972) hypothetical planet over an area extending at least $3\frac{1}{2}°$ from the predicted position in every direction. Two series of plates were taken with the Royal Greenwich Observatory's 33-cm Astrographic refractor with a minimum interval of one day between pairs of plates. The predicted daily motion of the planet across the plates was 0.7 mm. Each plate was blinked on a blink comparator down to at least magnitude 15.5 and in most cases fainter than magnitude 16.0. Foss, Shawe-Taylor, and Whitworth found no moving object brighter than the magnitude limit on any of the pairs of plates. They concluded that if a trans-Plutonian planet does exist it is either much less massive and hence

---

[13] Brady, J.L., The effect of a trans-Plutonian planet on Halley's comet. *Publ. Astron. Soc. Pac.*, 84, 314 (1972). By permission of the author and *Publications*; University of California, Lawrence Livermore Laboratory, and U.S. Department of Energy.

considerably fainter than Brady predicted, or else it is not near Brady's final position.

Hardie (1965a,b) made a new series of photoelectric measurements of Pluto in 1964 using the 60-cm Seyfert reflector at Dyer Observatory. He was able to refine the synodic period of rotation to 6.38673 days, or $6^d\ 9^h\ 16^m\ 54^s \pm 26^s$, which may differ from the sidereal or true rotation period by ±45 seconds. This uncertainty existed because the orientation of Pluto's axis of rotation and its direction of rotation were unknown. Hardie's data indicated that the mean brightness might have decreased a little from that measured in 1955. Hardie's (1965a) finding is contradicted by visual estimates of Pluto's brightness by Mosely (1969) who derived a mean visual magnitude of 14.0 from observations with the 25-cm refractor of Armaugh Observatory.

Hardie (1969) attributed the change in Pluto's magnitude to a change in the nature of its surface and hence albedo. Pluto is now nearer to the sun than its mean orbital radius of 5,864 million kilometers and Hardie suggested this would be sufficient to turn Pluto's possible covering of snow and ice made of frozen nitrogen to slush which would have a lower reflectivity. Such a phase change would explain Pluto's general dimming over the previous decade as well as its 6.39-day variations (due to a day/night-thaw/freeze cycle).

Lacis and Fix (1972) performed a Fourier analysis of the light curve of Pluto to obtain the longitudinal distribution of bright and dark areas on Pluto's surface. They found that although it was clear that a sizable difference in albedo existed between the brighter and darker areas, it could not be determined whether the light variation is due to dark spots or bright spots, or whether the spotted areas are large or relatively small. However, they suggested that if the two types of surface areas have different polarization characteristics and if changes in polarization with phase can be measured then the nature of Pluto's surface might be determined more reliably. They also pointed out that a spotted surface provides a better explanation for Pluto's light curve than either a non-spherical shape for Pluto or a highly tilted axis of rotation.

Andersson (1973), employing UBV observations of Pluto made in 1972 at McDonald Observatory and at the Lunar and Planetary Laboratory, calculated a mean magnitude equivalent to V =

15.15 at mean opposition distance compared to Walker and Hardie's (1955) original value of 14.90. Andersson gave an alternative explanation for this difference by suggesting that Pluto's rotational axis lies in the plane of the orbit; the axis was pointing towards the earth in 1930 when the planet was discovered. This model would require Pluto's polar cap to be about twice as reflective as its equatorial region.

Fix, Neff, and Kelsey (1970) measured the relative brightness of Pluto at 21 equally spaced wavelengths between 3400 and 5900 angstroms. Using these measurements and a normalized solar spectrum they determined the relative albedo of Pluto at each of the wavelengths (Fig. 5.7). They found that Pluto's albedo exhibits a general increase towards the red with a peak near 3800 angstoms and a definite depression around 4900 angstroms.

Manning (1971) noted, like Halliday (1969) earlier, that the nebular centrifuge theory requires that Pluto, because of its position in the solar system, not be an iron-rich planet. However he pointed out a significant general and specific resemblance between the reflectance spectrum of Pluto published by Fix, Neff, and Kelsey (1970) and the absorption spectrum of a terrestrial Fe-bearing silicate (Fig. 5.8). Manning noted, for example, the correlation between the 3780 angstrom depression in Pluto's spectrum and a prominent absorption feature near 3700 angstroms in the mineral spectrum. He concluded that there are sufficient correlations of energies and half-widths of bands in spectra of Pluto and terrestrial crystals to justify his suggestion that the surface, at least, of Pluto is iron-rich.

Webster, Webster, and Webster (1972) attempted a detection of Pluto at thermal wavelengths with the 26-m antennae interferometer of the National Radio Astronomy Observatory. They were unable to detect emissions from Pluto at either of the wavelengths, 11.1 cm (= 2.695 GH_z) and 3.7 cm (8.085 GH_z). The 8.085 GH_z observations yielded an upper limit of 162°K for the disk temperature of Pluto at the 3.7 cm wavelength.

According to Rubashevskii (1966) the first to suggest that it might be possible to establish the diameter of Pluto from occultation observations at several observatories was the Polish astronomer Banachiewicz, in 1941. Banachiewicz mistakenly estimated that Pluto would occult stars brighter than photographic magnitude 15 on the average of once a year and

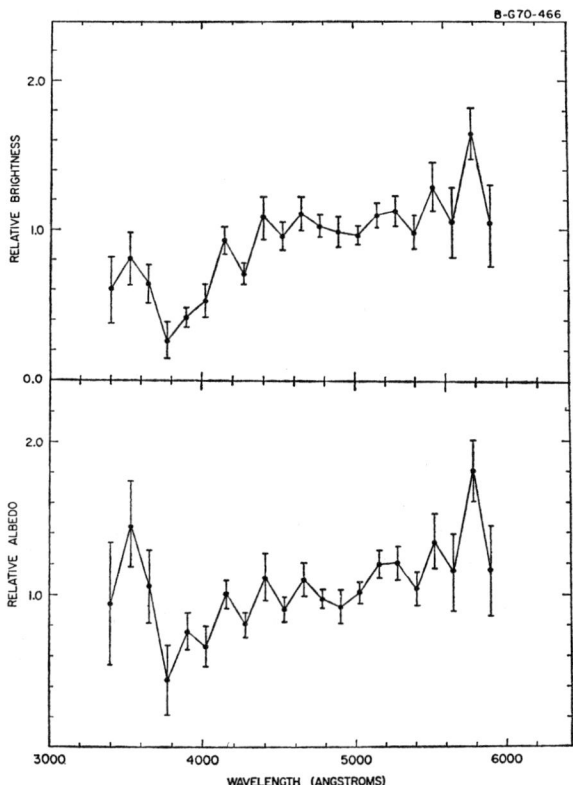

Fig. 5.7. The wavelength dependence of the relative brightness and the wavelength dependence of of the relative albedo of Pluto.[14]

---

[14]Fix, J.D., Neff, J.S., and Kelsey, L.A., Spectrophotometry of Pluto. *Astron. J.*, 75, 895 (1970). By permission of the authors and the editor of *Astronomical Journal*.

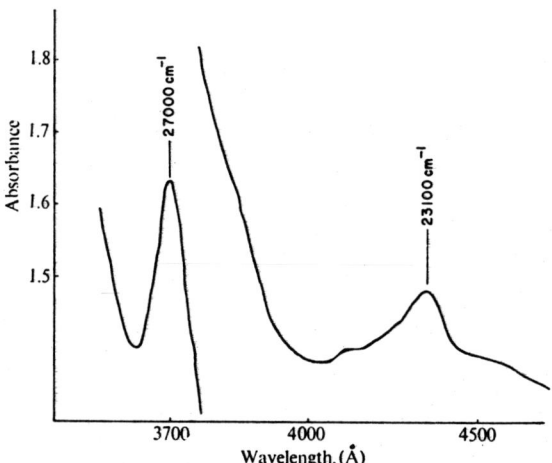

Fig. 5.8. Absorption spectra of terrestrial grossular garnet at 293 K[15]

---

[15]Manning, P.G., Is Pluto an iron-rich planet? *Nature*, 230, 234 (1971). With permission of the author and *Nature*.

suggested visual observations of these events would be feasible with telescopes of 30-cm to 50-cm aperture. Kordylewsky's (1956-58) unsuccessful attempts to detect any occultations led him to question the validity of Banachiewicz's estimate of occultation frequency.

Halliday (1963) made a proposal for a new determination of the diameter of Pluto by observing an occultation of a star at two or more observatories. He offered two possible explanations for the problem of reconciling the small diameter of Pluto as measured by Kuiper (1950) with the mass necessary to account for Pluto's supposed perturbations of Uranus and Neptune, the combination of which yields an unreasonable mean density for the planet of about $50 g/cm^3$. First he suggested the error might be attributable to the mass determination, which was necessarily based on old observations. If the mean density is assumed to be a more reasonable $4 g/cm^3$ then the mass of Pluto is only 0.07 times that of the earth - comparable to the size of Mercury or the largest satellites of Jupiter, Saturn and Neptune. Alternatively Halliday suggests the error might lay in the measurement of the diameter. Alter (1952) had earlier pointed out that the apparent diameter of Pluto may represent only a lower limit to the true diameter if the surface reflectivity is such that only the central portion appears illuminated when viewed from a great distance. Such limb darkening was, in fact, observed by Hardie (1965b). In considering the two alternatives, Halliday noted that an error in determining the diameter affects the mean density more critically than an error in the mass since the density depends on the cube of the diameter and only the first power of the mass.

Alter (1952) concluded that the only apparent hope of measuring the true diameter is to wait for Pluto to occult a star and even then he thought that many occultations would be required. After first determining that measuring the duration of an occultation should present no problems, Halliday (1963) suggested that a single occultation would suffice to obtain an accurate diameter for Pluto. This was because the earth and Pluto are of comparable size and thus the parallax involved in observing Pluto from two observatories can be a significant fraction of the angular diameter of Pluto which results in occultations of unequal length at the two observatories. The effect is quite sensitive for even a small separation of the two observatories. For occultations which are nearly central (along Pluto's diameter) the durations become nearly equal, but the very equality is proof of the fact they are nearly central and hence the effective diameter of Pluto will be measured by each observation.

The exact method is indicated in Fig. 5.9 where we suppose that a complete occultation has been observed by two observatories. First plot (on a greatly expanded scale) the path of Pluto in the sky as it would be seen from the centre of the earth. Mark the points A, B, C, D, along this path, where A and D are the positions at the times of disappearance and reappearance of the star as observed from one observatory, and B and C the corresponding points for the second observatory. These positions may be interpolated from ephemeris data for Pluto. High accuracy is required for the *relative* positions of the points A, B, C, D, but the whole plot need not be positioned in the sky with great precision. Using the standard formulae as given by the *Astronomical Ephemeris*, compute the geocentric parallax in right ascension and declination for each observation and plot the apparent positions for the centre of Pluto (points 1, 2, 3, 4) corresponding to the geocentric positions (A, B, C, D). This allows for the latitude and longitude of the observatories as well as the rotation of the earth, and involves the known distance of Pluto from the earth and also the diameter of the earth. The duration of the occultation as seen from the first observatory (positions 1 and 4) is fairly long and corresponds to a more nearly central occultation than is observed from the second observatory (positions 2 and 3).

Let the radius of Pluto be R. Then four circles, each of radius R, centered at the points 1, 2, 3, 4, must all pass through the position of the occulted star. Denote the position of the star (λ, θ) and the four known centres by (x_1, y_1) etc. Then we may write four equations of the form:

$$(\lambda - x_1)^2 + (\theta - y_1)^2 = R^2.$$

By combining any three of the equations we may solve for λ, θ, and R, i.e. the position of the occulted star relative to Pluto and the diameter of Pluto are determined. The fourth equation serves as a check.

Note that in principle only three equations are required. Thus for two observatories a solution is obtained even if one disappearance or reappearance is lost due to clouds or other trouble. For three observatories, suitably located, three observations of a disappearance are sufficient for a solution.[16]

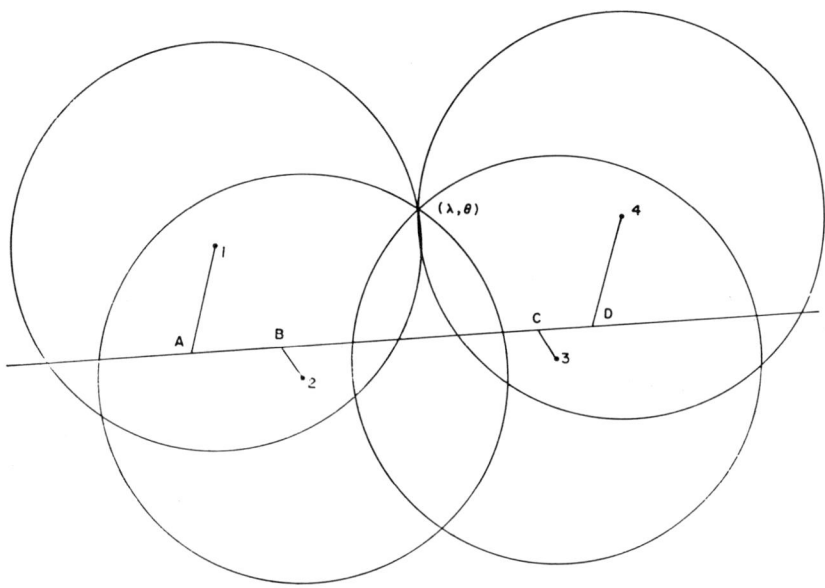

Fig. 5.9. Geometry of an occultation observed from two observatories, illustrating the effects of parallax.[17]

Halliday estimated the frequency of occultations to range from an average of 0.16 occultations per year for stars of visual magnitude 14 to 1.24 occultations per year of stars of visual magnitude 19 although these values would have to be reduced by 2 or 3 to take account of unfavourable planetary positions and weather conditions. Nonetheless, Halliday concluded that suitable occultations would occur every few years and that in view of the desirability of obtaining an accurate diameter for Pluto it was worthwhile to undertake an occultation prediction service for participating observatories.

[16,17]Halliday, I., A proposal for a new determination of the diameter of Pluto. *J. R. Astron. Soc. Canada*, 57, 163 (1963). By permission of the author and the editor.

In the interval between Halliday's (1963) proposal and the first attempt at observing such an occultation two years later Hardie (1965b) published data indicating a significantly larger diameter for Pluto than the 5900 kilometer value established by Kuiper (1950). This was based upon photometric observations carried out in 1964 which produced a light curve of essentially the same shape as but possibly of slightly larger amplitude than that found earlier (Walker and Hardie, 1955). The light curve shows that Pluto's light increases for about four days then drops in about two days during each rotation period now refined to $6^d\ 9^h\ 16^m\ 54^s \pm 26^s$. This indicates that Pluto is brightest at its centre with appreciable limb darkening.

Following Halliday's proposal a program was undertaken to measure the path of Pluto among the faint background stars on plates taken with the 120-cm Schmidt telescope at Palomar Observatory and the 60-cm Seyfert telescope of the Dyer Observatory. Stellar positions were measured on the Mann blink comparator at the Dominion Observatory at Ottawa. A short computer program was then used to determine the minimum separation between Pluto and any star lying close to its predicted path, together with the time of closest approach. A possible occultation of a star of visual magnitude 15.3 was predicted for the night of April 28/29, 1965 (Halliday, 1965a). Halliday, in announcing the predicted occultation, indicated that the combined light of the star, located at $\alpha = 11^h\ 23^m\ 12^s.1$, $\delta = +19°\ 47'\ 32"$ (1950), and Pluto would be magnitude 13.8. Thus if the occultation actually occurred the magnitude would suddenly drop to Pluto's 14.1 and the resulting 25 percent drop in light should be observable in telescopes with apertures over 50-cm. Observatories participating in this attempt to time the expected occultation included virtually all those in North America possessing telescopes of adequate aperture and suitable photometers ranging in latitude from Victoria, British Columbia to Fort Davies, Texas.

Unfortunately no occultation was observed: the upper edge of the occultation track passed just south of the southernmost participating observatory, McDonald. However, in their summary reports Halliday (1965b) and Halliday, Hardie, Franz, and Priser (1966) found that it was possible to derive an extreme upper limit of 6800 kilometers for Pluto's diameter. The results of this observation program were presented in the 1966 paper by the following diagram (Fig. 5.10):

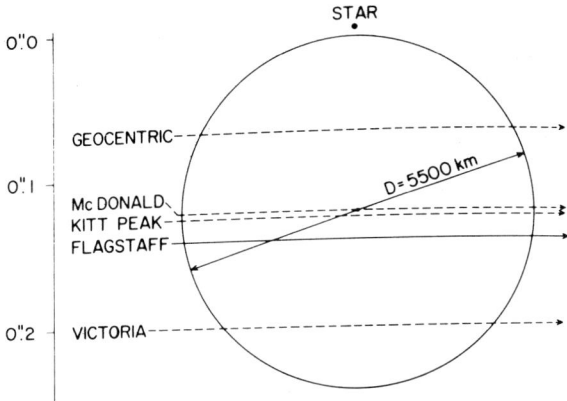

Fig. 5.10. A representation of photocentric paths of Pluto past the star.[18]

The star's position is indicated by the dot; the solid line represents the photocentric path of Pluto at a distance of 0".143 from the star. The minimum photocentric separation for the McDonald Observatory is reduced to 0".125, corresponding to a diameter of Pluto of 5800 kilometers for a grazing occultation. The circle represents the projected disk of the planet as seen from McDonald if Pluto were 5500 kilometers in diameter. Allowing for an estimated mean error of 0".013, one can assign an extreme upper limit of 6800 kilometers to the diameter. Thus Kuiper's value of 5900 kilometers, while not directly confirmed, falls between the new upper limit of 6800 kilometers and the lower limit of 2000 kilometers; the latter diameter corresponding to unit albedo. The upper limit of 6800 kilometers corresponds to an extreme lower limit for Pluto's albedo of 0.1. Combining a diameter of 6800 kilometers with the highest reasonable density, $5.5 g/cm^3$ (that of earth), results in a revised upper limit for Pluto's mass of 0.14 earth masses, or 2,200,000 reciprocal solar masses. Even this highest value for the mass of Pluto is insufficient to account for the perturbations in the motions of Uranus and Neptune that were used in deriving the mass and predicting

---

[18]Halliday, I., Hardie, R.H., Franz, O.G., and Priser, J.B., An upper limit for the diameter of Pluto. *Publ. Astron. Soc. Pac.*, 78, 113 (1966). By permission of the authors and *Publications*.

the existence of Pluto. Halliday and his colleagues (1966) pointed out that since the diameter of Pluto was now known with confidence to be considerably smaller than that of the earth, the parallax effect described above would be capable of even greater accuracy, and that the diameter of Pluto may ultimately be known to within one percent of its true value. Sanders (1965) published results of the occultation observation attempts at Lick Observatory with the 90-cm refractor and 300-cm reflector.

Rubashevskii (1966) outlined an occultation observation program similar to Halliday's (1963), differing primarily in the method of predicting occultations by Pluto, and containing somewhat more pessimistic estimates of the frequency of such occultations. He pointed out the desirability of observing occultations by Pluto from balloons or satellites, apparently in anticipation of such earth-orbiting observatories as the 240-cm Space Telescope that will be placed in orbit in late 1983 by the Space Shuttle. He noted that it was now evident, as a result of Halliday's (1965b) work, that the source of the contradiction between the mass and the dimensions of Pluto lay in a very inaccurate determination of the planet's mass from perturbation theory, and not in the "quasi-specular" character of the reflection of sunlight from its surface as had been earlier believed (Hardie, 1965b). Rubashevskii concluded that it might very well be that the discordance with the mass of Pluto arises from an omission of the effects on its motion produced by a hypothetical trans-Plutonian planet.

O'Leary (1972) urged that a systematic and expanded system of occultation predictions be initiated to include, in the case of Pluto, stars as faint as photographic magnitude 18. He noted that in view of the short lead times (of the order of a few weeks) for the fainter and more frequent events, a system of cooperation among observatories with suitable instruments was needed. He estimated that the number of "passable" occultations by Pluto, defined as those involving a light intensity drop of greater than ten percent, was at least one event per year. It is interesting to note that Barbieri, Capaccioli, and Pinto (1975) report a near-occultation of a 17th magnitude star by Pluto on a plate taken with the Asiago Schmidt telescope on May 25, 1974.

Chapter 6
Cold Moonlight
1973-1979

Andersson and Fix (1973) presented an analysis of new photometric observations of Pluto's light variation made in 1971, 1972, and 1973. They found, as had Walker and Hardie (1955) and Hardie (1965b) earlier, no measurable variation in Pluto's color with rotational phase. They compared the three light curves obtained from observations made in 1954-1955 (Walker and Hardie, 1955), 1964 (Hardie, 1965b), and 1973. They interpreted the gradual decrease in the mean brightness and increase in the amplitude during the past few decades (Fig. 6.1) as being caused by a change in the aspect of the planet due to its orbital motion and the large obliquity of the planet's axis of rotation.

Elaborating on Andersson's (1973) earlier suggestion, Andersson and Fix (1973) attempted to determine the orientation of Pluto's rotational axis through an analysis of the photometric data. For this analysis they followed Russell's (1906) discussion of the general problem of interpreting the light curve of a rotating body in terms of its surface brightness distribution. They expressed the light curve as:

$$L = K_0 + K_1 \cos(\omega t + \phi_1) + K_2 \cos(2\omega t + \phi_2) + \ldots, \quad (1)$$

where ω is Pluto's synodic rotation rate, t is time, and K_1 and ϕ_1 are constant coefficients. The Fourier coefficients K_1 are themselves functions of the surface albedo distribution and θ, the angle between Pluto's axis of rotation and the line of sight. K_0 is Pluto's mean brightness. K_1 is a measure of the amplitude of Pluto's light curve, and K_2 is a description of the light curve's asymmetry. Andersson and Fix tabulated the mean magnitudes and amplitudes of Pluto during several observational periods (Table 16).

The magnitudes given in Table 16 are reduced to mean oppo-

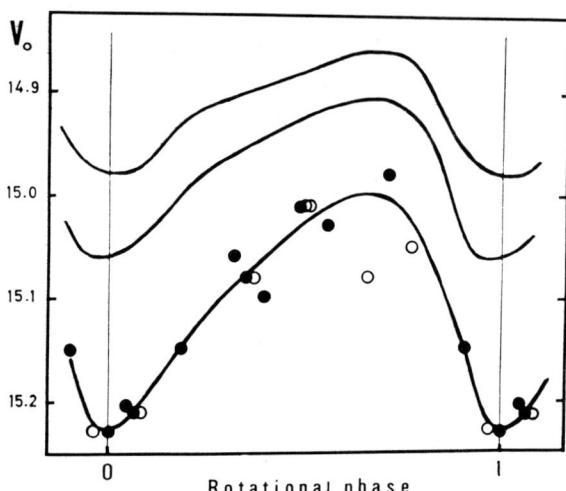

Fig. 6.1. Pluto light curves 1954-1955 (top), 1964 (centre) and 1972 (bottom).

The light curves are reduced to mean opposition and a solar phase angle of 1°. The period used is 6.3867 days. Individual observations in 1971 - 1973 are plotted; observations of low weight are shown as open circles, others as filled circles.[1]

---

[1] Andersson, L.E., and Fix, J.D., Pluto: New photometry and a determination of the axis of rotation. *Icarus*, 20, 279 (1973). By permission of the authors and *Icarus*. (Copyright [1973] by Academic Press, Inc.).

TABLE 16. Mean Magnitude and Amplitude of Pluto[2]

Date	V	Amplitude	Reference
1930-33	14.75:	-	Graff (1930), Nicholson and Mayall (1930), Baade (1931)
1953.3	14.92:	-	Walker and Hardie (1955)
1954.2	14.92	0.11	Walker and Hardie (1955)
1955.3	14.88	.11	Walker and Hardie (1955)
1964.4	14.99	.15	Hardie (1973*)
1966.3	14.90	-	Kiladze (1967)
1972.0 / 1972.4	15.12	.22	Andersson and Fix (1973)

*Private communication to Andersson and Fix.

---

[2] Andersson, L.E., and Fix, J.D., Pluto: New photometry and a determination of the axis of rotation. *Icarus*, 20, 279 (1973). By permission of the authors and *Icarus*. (Copyright [1973] by Academic Press, Inc.).

sition and to a solar phase angle of 1° using a phase coefficient of 0.05 mag/deg. The magnitude given for 1930-33 is an average of the visual and photographic estimates (converted to V magnitudes) made by several workers shortly after Pluto's discovery. Andersson and Fix attribute the difference in V_0 between 1954 and 1955 mainly to calibration errors. They were unable to explain the apparent disagreement between Kiladze's (1967) observations in 1966 and the trend of the other V_0 data.

The value of θ for a given observation is determined by the orientation of Pluto's pole. Accordingly, Andersson and Fix searched the right ascension, declination plane for the locations of those poles which gave, in terms of the adopted model, satisfactory agreement with the existing photometric data. The goodness of fit (found by least squares) requirements for K_0 and K_2 proved to be very unrestrictive so that the domain of possible pole positions was determined almost entirely by the goodness of fit to the K_1 coefficients and the matching of the early observations of Pluto's mean brightness. Owing to the limited precision of the light curves, the calculations were repeated using various combinations of K_1 differing from the original values by as much as 5 percent for K_1 and K_2 and 3 percent for K_0. It was found in all cases that the domains of possible pole positions lay close to the domain found by the original search. The region encompassing the various domains is shown in Fig. 6.2.

There is, of course, another domain of pole positions lying 180° from the domain shown in Fig. 6.2 and Andersson and Fix were unable to state whether the pole indicated was a north or south pole. They noted that it was apparent Pluto's obliquity is large, probably greater than 50°. They commented that it would be very informative to repeat their calculations after the light curve of Pluto had been measured again in the late 1970's. Much of the domain of possible pole positions shown in Fig. 6.2 would prove to be incompatible with the new light curve and it should thus be possible to determine the coordinates of Pluto's pole with considerably increased acccuracy.

Kelsey and Fix (1973) reported on the results of polarimetric observations of Pluto involving measurements of the linear polarization of the light reflected by the full visible disk of Pluto. Several observations were made on each of six nights in April, 1972 using a polarimeter calibrated against three comparison stars and attached to the 1.3-m telescope of the Kitt Peak National Observatory. For each of these obser-

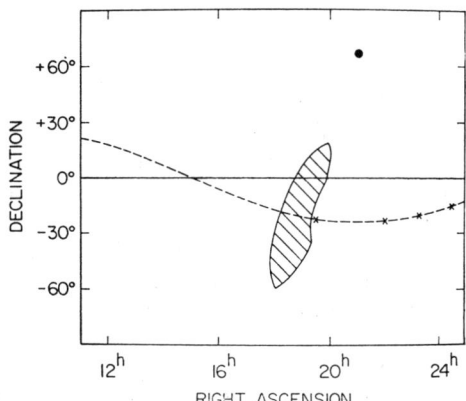

Fig. 6.2. The domain of possible pole positions.

    The orbit of the Sun, as seen from Pluto, is indicated by the broken line. The positions of the Earth and Sun for 1932, 1954-1955, 1964, and 1972 are shown, from left to right, by the X's. The pole of Pluto's orbit is indicated by the filled circle.[3]

---

[3]Andersson, L.E., and Fix, J.D., Pluto: New photometry and a determination of the axis of rotation. *Icarus*, 20, 279 (1973). By permission of the authors and *Icarus*. (Copyright [1973] by Academic Press, Inc.).

vations Kelsey and Fix determined the degree of polarization (P) and its probable error, both in percent; and the position angle (θ) and its probable error, both in degrees according to the 6.3867-day light curve of Pluto. The mean degree of polarization was 0.27 percent with a probable error of 0.02 percent, and the mean position angle was 156° with a probable error of 2°. At the time of the observations Pluto had a solar phase angle of 0°.8.

In view of Pluto's small, albeit uncertain, geometric albedo, it was not unexpected that Pluto showed a relatively large degree of polarization at such a small phase angle. Although extrapolation from polarization at 0°.8 phase angle to minimum polarization (P_{min}) in the vicinity of phase angle 10° could not be done with any precision, it seemed clear to Kelsey and Fix that P_{min} for Pluto is at least 1 percent. The albedo-P_{min} relation for airless bodies (Veverka, 1971a,b,c and Dollfus, 1961, 1971) thus implied that the albedo of Pluto is less than about 0.25, a limit consistent with values of Pluto's albedo calculated from its brightness (Harris, 1961) and radius (Kuiper, 1950 and Halliday, Hardie, Franz and Priser, 1966). However, the assumption of an airless-body state for Pluto may be untenable (see Hart, 1974 and Golitsyn, 1975). The polarization curve of Pluto would appear to indicate a rough surface with an extensive region of meteoroid bombardment.

Kelsey and Fix also examined the polarization data for evidence of variability associated with the rotation of Pluto; but they found none, probably because the amount of change in the degree of polarization was indistinguishable in their data. However, they predicted that observations made when Pluto's solar phase angle is near its maximum and Pluto's degree of polarization is accordingly higher should permit determination of whether Pluto's brightness variations are associated with polarization changes. When viewed at small phase angles, the position angle of polarization of a smooth-surfaced planet would be equal to the position angle of the intensity equator of the planet. Figure 6.3 shows the nightly position angles of polarization of Pluto as well as the position angle of Pluto's intensity equator during the time of the observations made by Kelsey and Fix, who point out that the measured position angle for Pluto is systematically greater than that predicted for a planet with a featureless surface.

Newburn and Gulkis (1973) in a survey of the outer planets and

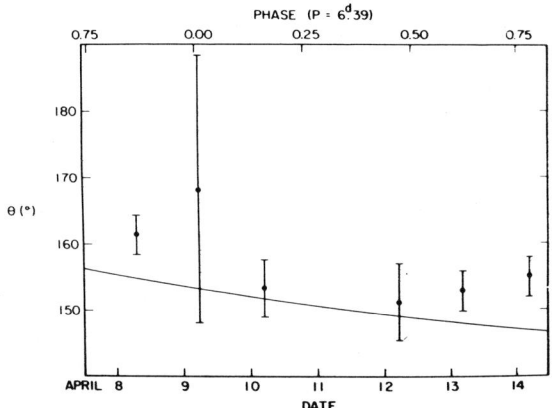

Fig. 6.3. The nightly values of the position angle of polarization for Pluto versus date and versus phase for a period of 6.39 days.

The solid line is the position angle of the intensity equator of Pluto during the time of the observations.[4]

---

[4]Kelsey, L.A., and Fix, J.D., Polarimetry of Pluto. *Astrophys. J.*, 184, 633 (1973). Reprinted courtesy of the authors and *The Astrophysical Journal*, published by the University of Chicago Press. (Copyright [1973] by American Astronomical Society).

their satellites reviewed the then current state of knowledge concerning Pluto. They noted the high eccentricity and inclination of Pluto's orbit and also its stability as established by Cohen and Hubbard (1965) and Williams and Benson (1971). Such stability would seem to weaken the case for the escaped satellite of Neptune hypothesis. Newburn and Gulkis reiterated the large uncertainties in existing estimates of Pluto's diameter and mass. They concurred with Ash, Shapiro, and Smith (1971) that "Pluto's mass cannot be determined reliably from existing data." They suggested that only a spacecraft flypast will allow an accurate mass determination any time in the near future. They also suggested that a spacecraft occultation experiment as being the best means of detecting an atmosphere of Pluto. However, they observed that Pluto may not have an atmosphere because of its low mass and temperature. Many potential atmospheric constituents such as CO_2, H_2O, and NH_3 would be frozen out while others such as H_2 and He may well have escaped. The heavier inert gases such as neon and argon could form a permanent atmosphere, if present, but they would be very difficult to detect spectroscopically from Earth. Newburn and Gulkis noted that because Pluto's radius was so indefinite, "attempts to quote a geometric albedo are almost meaningless." With even the density very uncertain, they suggested there are not likely to be any models for Pluto's interior until many measurements are made from a spacecraft. A magnetic field was deemed unlikely since Pluto is small and rotates slowly.

Kiang (1973) examined Brady's (1972) proposition that a trans-Plutonian planet is disturbing the motion of Halley's comet. He set out to show that rather than an unknown planet, it is the non-gravitational force envisioned by Whipple (1950) that is the disturbing agency. After briefly acknowledging the possibility of a non-gravitational explanation for the residuals in the motion of Halley's comet, Brady had then shown how the residuals practically vanish when an empirical term, ε_6, is incorporated in the calculations of the comet's orbit. Kiang observed that although the achievement of this ε_6 term is impressive, it was brought into the calculation *ad hoc* and has no physical interpretation. Kiang pointed out that the term is mathematically equivalent to the anomalous deceleration of Halley's comet found in a previous paper (Kiang, 1971) and furthermore, the mathematical equivalent of the term can be given a plausible physical interpretation. Kiang (1973) showed that the size of the non-gravitational force involved is quite comparable to those found in other comets

(Marsden, 1969, 1970, and 1971 and Yeomans, 1971). Brady (1972) had instead attempted to find a planetary orbit and mass beyond Pluto to include in the perturbations of Halley's comet which would do as well as the secular term, ε_6, in reducing the residuals. Kiang (1973) expressed skepticism at the object, having a mass almost equal to that of Jupiter and moving in a highly inclined orbit in the retrograde sense, that Brady was eventually led to postulate.

Rawlins and Hammerton (1973) performed an analysis of the residuals of Neptune to delimit the range of possible masses and positions for a hypothetical tenth planet of the solar system. They reviewed the difficulties encountered in any study of the disturbances caused by an unknown outer planet, including the large uncertainties in the masses of some of the known planets. Therefore they decided to base each mass on the motion of the outermost well-observed satellite, *i.e.* Saturn: 1/3494 (from Iapetus) and Uranus: 1/22685 (from Oberon). However, they noted, Pluto (then) had no observed satellite and Neptune's mass derived from its satellite, Nereid, disagreed with that from other estimates. Therefore Rawlins and Hammerton set the mass of Neptune as an extra unknown when analyzing Uranus's longitudes. They performed a differential analysis of residuals, employing classical perturbation theory involving a rigorous least-squares fit.

Neptune was used as the primary source of data, although data from Uranus's longitude were later incorporated as a control and the hypothetical planet's orbit was assumed to be circular. Rawlins and Hammerton obtained limits of solutions for longitude and orbital radius and then latitude solutions for several values of longitude and radius. They also provided a table of probable upper limits on planets in circular orbits in the plane of the ecliptic for various combinations of longitude from 0° to 360° and orbital radius from 43 a.u. to 600 a.u. They noted that any planet beyond 600 a.u. would probably have been set askew or adrift long ago by passing stars.

Rawlins and Hammerton then made a sample comparison of observed and calculated longitudes for the close-fit position, $\alpha = 0.5$, $\varepsilon' = 330°$. Their postulated planet accounted well for Lalande's 1795 observation and caused only slight disturbances in the post-discovery Neptune data used. However, they conceded that this was because the hypothetical planet was very far from Neptune for most of that time. Yet, they point out, it is unlikely that Lalande should twice have made errors of 7". They also noted that the planet, which they suggested should not be dimmer than magnitude 17, would likely have been of low declination during Tombaugh's (1961) survey, thus making its possible discovery more difficult.

Neff, Lane, and Fix (1974) analyzed new photometric observations of Pluto made with the 90-cm reflector at Kitt Peak National Observatory on four consecutive nights in April, 1973 in an attempt to resolve the long-standing ambiguity in Pluto's period of rotation. Although most workers, including Walker and Hardie (1955) had interpreted the photometric data as indicative of a 6.39-day period, a shorter period of 1.1819 days was also tenable. Neff and his colleagues found that the observed and predicted mean magnitude for the short period differed significantly on one night and the observed and predicted slopes of a plot of magnitude *versus* phase differed significantly on three nights. Four of the observed magnitudes and three of the observed slopes were in good agreement with predicted ones for the longer period in a plot of identical ordinate and abscissa. On two nights the observations had the same phase on the short period and the magnitudes were found to differ by $0^m.18$, about eight times the average standard deviation. Neff and his colleagues therefore concluded that the 1.1819-day period was spurious and that the 6.3867-day period, which they were able to refine to 6.3874±0.0002 days, was confirmed.

Cohen and Hubbard (1965) had shown that the minimum separation of Pluto and Neptune is limited to 18 a.u. because of the resonance arising from the near commensurability of the motions of Pluto and Neptune in the ratio of two to three. Brouwer (1966), in discussing the results of Cohen and Hubbard, commented that before valid conclusions could be drawn concerning the stability of the Pluto-Neptune system over a much longer period of time, the long-term motion of the argument of perihelion must be investigated. This is because if the perihelion of Pluto continues to progress in a direct orbital sense, the radius vector of Pluto at perihelion passage would occasionally coincide with the orbital planes of the other outer planets with a consequent possibility of closer approaches to those planets and a very long-term alteration of the motion of Pluto. However, if the perihelion of Pluto were to librate about a mean value so that the perihelion were to remain away from the planes of Neptune and the other planets, then the stability of the system would be enhanced and the avoidance of close encounters assured. Williams and Benson (1971) provided just such a long-term investigation of Pluto's motion for a period of 4.5 million years.

Nacozy and Diehl (1974) further explained and confirmed the stable nature of the librational motion of the perihelion of Pluto found by Williams and Benson. Nacozy and Diehl found that the eccentricity (e) and inclination (i) of Pluto, and two angular coordinates, the first critical argument (δ_1) and the argument of perihelion (ω), exhibit long-term libration about stationary values. In general, they note, libration of a coordinate about a stationary value indicates stability for that coordinate. Since the actual orbit of Pluto appears to have a stable libration about a stationary point, Nacozy and Diehl suggested that the 4.5 million year solution of Williams and Benson may be extrapolated to much longer periods in the past and future.

Baldi and Caputo (1974) suggested that from the planets whose fundamental parameters: mass (M), equatorial radius (R), rotation period (w), and J_2 (the second term of the gravitational potential) are known, we can obtain interesting information regarding the moments of inertia (C) of Uranus and Pluto and the mass of Pluto. They found that their calculated value for the density of rotational angular momentum (Cw/M; [cm^2/s]) for Uranus fit very well the straight line obtained by MacDonald (1964) in his plot of Cw/M versus M for planets whose moments of inertia were known with a fair degree of precision. Baldi and Caputo then tested the goodness of fit to MacDonald's straight line of seven combinations of Cw/M and M derived from radii of 3200 km (Halliday, Hardie, Franz, and Priser, 1966) and 2700 km (Halliday, 1969) and mass values ranging from 0.11 earth masses (Seidelmann, Klepczynski, Duncombe, and Jackson, 1971) to 0.004 e.m. with the period of rotation being 6.387 days. They found that the only reasonable models of Pluto were those involving a radius of 2700 km and a mass range of 0.004 to 0.0138 earth masses and a radius of 3200 km and a mass range of 0.0069 to 0.023 e.m. The other three models were either too far above or too far below the straight line of MacDonald's plot of Cw/M versus M.

Horedt (1974) did a numerical integration of the plane circular restricted three-body problem in the case of an exponential mass variation of the small primary m (i.e. planet) and for a satellite of m. He found that for a mass loss by a factor of twenty the satellite escapes towards the regions inside the orbit of m, provided that the mass loss rate is sufficiently slow. Extrapolating his results to the real solar system, Horedt concluded that any satellite escaping due to a considerable mass loss by proto-Neptune would move after escape very likely inside the orbit of Neptune, and not outside it as would be required for Pluto to have originated in this manner. Therefore, it seemed probable that Pluto originated as an independent trans-Neptunian planet from the protoplanetary nebula.

Hart (1974) notes that spectroscopic observations of Pluto have not yet revealed the presence of any atmosphere, supporting the view that all common gases have either frozen out at Pluto's very low average temperature (~43°K) or escaped from its weak gravity. He points out, however, that neon has a high enough molecular weight to avoid escaping and is volatile enough to avoid freezing out. Therefore, Hart suggested that Pluto may possess an atmosphere of neon. Such an atmosphere, even if quite thick, would not produce any absorption lines in the visible region of the spectrum.

Adopting a mass value (M) for Pluto of 0.11 earth masses (Seidelmann, Klepczynski, Duncombe, and Jackson, 1971), a radius (R) of 2900 km (Kuiper, 1950), an eccentricity (e) of 0.246 (Cohen, Hubbard, and Oesterwinter, 1967), a semi-major axis (a) of 39.53 a.u. (*op. cit.*), and a Bond albedo (A) of 0.17 (Kuiper, 1950), Hart calculated the amount of sunlight absorbed by Pluto at any given time. From this, and assuming Pluto radiates at any instant as much energy as it absorbs, Hart computed its effective temperature at various distances from the Sun (Table 17). The second column of Table 17 gives the effective temperatures assuming the day and night-side temperatures of Pluto to be equal. The third column in Table 17 lists the effective temperature of the day-side assuming the night-side to be much colder.

The vapour pressure of liquid neon, as a function of temperature, is given in Table 18. The triple point of neon is at: T = 24.5 K, P = 0.427 atm. Its critical point is at: T = 44.4 K, P = 26.9 atm. Assuming equal day and night-side temperatures on Pluto, joint inspection of Tables 17 and 18 shows that neon will not freeze out. If the surface pressure of Pluto should be high enough (about 21.5 atm. at a temperature of 42.9 K) then liquid neon will exist at the surface.

It is likely that neon will be the only gas present in any quantity. Most common gases such as carbon dioxide, ammonia,

TABLE 17. Effective Temperatures[5]

Pluto's distance from Sun	Daytime temperature, in K, assuming	
	(a) $T_{night}=T_{day}$	(b) $T_{night}=0°K$
$a(1-e)$	49.0	58.3
$a(1-e^2)^{1/4}$	42.9	51.0
a	42.5	50.6
$a(1+e)$	38.1	45.3

TABLE 18. Vapour Pressure of Liquid Neon[6]

Temperature (K)	Pressure (atm)
30	2.2
31	2.7
32	3.35
33	4.1
34	5.1
35	6.2
36	7.5
37	9.0
38	10.8
39	12.5
40	14.5
41	16.8
42	19.5
43	21.9
44	24.5
44.4	26.9

[5,6] Hart, M.H., A possible atmosphere for Pluto. *Icarus*, 21, 242 (1974). By permission of the author and *Icarus*. (Copyright [1974] by Academic Press, Inc.).

and methane are solids at even the highest temperatures which
might exist on Pluto. The second most abundant gas is likely
argon; but even at a temperature of 58°K argon has a vapour
pressure of less than 0.01 atm. Even if there is any free,
uncombined nitrogen present, its vapour pressure is only about
0.04 atm at 58°K and even less at lower temperatures. Most of
the hydrogen and helium originally present will probably have
escaped from Pluto long ago. Hart calculated that whether
Pluto's atmosphere is isothermal or not, the planet could retain an atmosphere of neon without difficulty. The mass (M_{Ne})
of its neon atmosphere would be the main parameter determining
conditions on the surface of Pluto. Table 19 presents estimates of some important quantities as a function of M_{Ne}. It
will be seen that if M_{Ne} is as great as 4.61×10^{22} g, oceans
of liquid neon will cover all or part of the surface of Pluto.
If neon is slightly less abundant oceans would exist part of
the year, drying up completely during the "hot" season when
Pluto was nearest the Sun. If M_{Ne} is less than 4.46×10^{22} g
oceans will not exist at all.

An atmosphere massive enough to form oceans would, because of
higher reflectivity in the shorter wavelengths due to Rayleigh
scattering, cause the albedo of Pluto to be much higher in the
ultraviolet than in the red. However Fix, Neff, and Kelsey
(1970) found exactly the opposite trend in Pluto's albedo.
Comparison of theoretical values of monochromatic geometric
albedo for different values of atmospheric pressure with observations of Pluto's albedo shows that although the data are
compatible with a surface pressure as large as 1.0 atm, a surface pressure as great as 3.0 atm seems unlikely. Therefore
Hart concluded that neon oceans should not occur at anytime on
Pluto.

Golitsyn (1975) examined the thermal and dynamical properties
of the type of neon atmosphere ($T_{average} \simeq 43°K$, $P \simeq 1.0$ atm) proposed by Hart. He calculated that the velocity of sound waves
would be 170 m/sec and the characteristic cooling time of the
atmosphere would be about 200 years. Since Pluto revolves
about the Sun in 248.6 years, the seasonal temperature variations induced by its large orbital eccentricity ($e = 0.246$)
and the possible large obliquity of its axis of rotation
(Andersson and Fix, 1973) should be strongly damped for such
an atmosphere. Hence, Golitsyn suggested, small variations in
the temperature of Pluto as it moves in its orbit could serve
as evidence of a substantial atmosphere. The characteristic
temperatures, δT, inducing atmospheric circulation and the

TABLE 19. Some Effects of a Neon Atmosphere[7]

M_{Ne} (10^{20} g)	M_{Ne}/M[a] (ppm)	Surface Pressure (atm)	Are there neon oceans?	$T_{day}-T_{night}$ (K)	$\Delta(T_{eff})$ (Seasonal) (K)
<2.14	<0.32	<0.1	Never	Possibly large	<11.0
2.14-446	0.32-67.8	0.1-20.9	Never	<2.0	<11.0
446-461	67.8-70.1	20.9-21.5	Part of the year	<0.4	<0.6
>461	>70.1	21.5	All year	<0.4	<0.06

[a]For the Sun this figure ~ 1000 ppm.

[7]Hart, M.H., A possible atmosphere for Pluto. *Icarus*, 21, 242 (1974). By permission of the author and *Icarus*. (Copyright [1974] by Academic Press, Inc.).

wind velocity, U, were calculated to be $\simeq 0.07°K$ and $\simeq 30$ cm/sec respectively. Such slight temperature differentials are a consequence of the large thermal inertia of the atmosphere at low temperatures which ensures effective heat transfer into regions of the planet poorly warmed by the Sun despite the low wind velocities. Golitsyn suggested the atmospheric circulation would be symmetrical with a single direct circulation cell extending from the equator to the pole.

Golitsyn said that the low value of Pluto's albedo excludes the presence on the planet's surface of a dense frost composed of, for example, argon or molecular nitrogen. This fact enables us to set upper limits on the abundance of these substances in the atmosphere of Pluto. He suggested that in order for there not to be polar caps at the poles of Pluto, where $T \simeq 42°K$, the partial pressure of nitrogen should be less than 0.4 mb. Even if it is assumed that the 20 percent decrease in albedo observed (Andersson and Fix, 1973) during the 1955-1972 period occurred because of a thaw pre-existing patches of nitrogen frost on the surface of the planet as it approached the Sun (rather than because the orbital motion of Pluto presented darker equatorial regions of the planet towards the Earth), an upper limit on the molecular nitrogen abundance in the atmosphere would still be 1 mb. Yet, even so rarefied an atmosphere would completely modulate the thermal conditions on Pluto's surface, and in principle it might be detectable from radio emissions by the planet (see Webster, Webster, and Webster, 1972).

Hart (1974) conjectured that neon was the only gas likely to form an atmosphere on Pluto. Other gases such as hydrogen and helium would likely have escaped while others such as carbon dioxide, ammonia, and methane may have formed frost on the surface of Pluto. Golitsyn (1975) interpreted the low geometric albedo of Pluto as indicative of the absence of such frost. However, Cruikshank, Pilcher, and Morrison (1976) later reported finding evidence of methane frost on Pluto, suggesting that its albedo may be higher and its diameter less than previously estimated.

Cruikshank and his colleagues designed a simple observational test to distinguish between the ices or frosts of water, ammonia and methane. The presence of frost on a planetary surface is strongly indicated by a generally decreasing reflectance with increasing wavelength between 1 and 4 µm. Cruikshank noted that reflectance measurements between 1.4 and

1.9 μm can distinguish CH_4 from either H_2O or NH_3. Methane shows a deep absorption at 1.7 μm, while the other two frosts show reflectance maxima near this wavelength and absorptions at 1.5 μm instead. Water and ammonia can in turn be distinguished by reflectance measurements between 3.2 and 3.6 μm, where H_2O is virtually non-reflective and NH_3 is highly reflective.

Cruikshank and his co-workers observed Pluto with the 4-meter reflector at Kitt Peak National Observatory in March 1976, measuring the planet's reflectance with standard *JHK* filters as well as with two specially made narrow filters centered at 1.55 and 1.73 μm (designated *H*1 and *H*2, respectively). The measured broadband colors for Pluto, expressed as magnitude differences, were *J* - *H* = 0.2±0.1 and *H* - *K* = -0.4±0.1, with the latter being about that expected for a frost while the former is considerably more positive than that expected for a water frost-covered object. The *H*1/*H*2 reflectance ratio for Pluto is inconsistent with water frost, being 1.6±0.1 compared with ~ 0.5 for laboratory spectra of H_2O frost.

On the basis of Pluto's *H*1/*H*2 ratio, its *J* - *H* color, and other observational and theoretical considerations cited in their paper (1976), Cruikshank, Pilcher, and Morrison concluded that CH_4 frost is probably the dominant reflecting material on Pluto's surface. They suggest that the *H*1/*H*2 ratio for Pluto is less than that for a pure laboratory sample of CH_4 frost because Pluto's surface is either mixed with dust or other frozen volatiles as suggested by Kuiper (1950), or has a different grain size distribution than the laboratory sample. They also note that the 20 percent brightness variation exhibited by Pluto as it rotates indicates the frost cover is not uniform. They present two plausible explanations for the origin of Pluto's CH_4 covering, the most straightforward being that the temperature of the proto-planetary nebula dropped below the freezing point of CH_4 at Pluto's distance from the Sun. The other possibility is that the temperature dropped only to the point of condensation of the hydrate, $CH_4 \cdot 8H_2O$, near 70°K. This hydrate would have subsequently dissociated to leave a surface coating of more or less pure CH_4 ice.

Cruikshank and his colleagues noted that the previously assumed geometric albedo for Pluto of about 0.1 would have to be revised upward. Taking into account the likely patchiness of the frost covering they suggest a revised average geometric albedo of 0.4 which implies a new diameter for Pluto of 3300 km, slightly smaller than that of our moon. An albedo of 0.6 would imply a diameter of only 2800 km. Such a small size, coupled with reasonable mean density values, yields mass values for Pluto much too small for it to have been the source of the residuals in the motions of Uranus and Neptune. They concluded that Tombaugh's (1961) discovery of Pluto in 1930 was thus the result of the thoroughness of the search rather than the validity of the predictions of its existence.

The *UBVRI* (0.37 to 0.82 µm) photometric measurements of Pluto made by Hardie and reported by Harris (1961) can be used to compare Pluto's reflectance with that of other objects in the solar system. Lane, Neff, and Fix (1976) noted, however, that Hardie's *I* filter was located at a wavelength just shortward of the blue side of the 1.0 µm absorption feature common to many asteroids and meteorites, so that a measurement of Pluto's reflectivity at a slightly longer wavelength than Hardie's *I* filter would be valuable for comparing Pluto with these other objects. Lane and his colleagues therefore measured Pluto's relative reflectance with an optical interference, high-pass infrared filter with an effective wavelength of 0.86 µm. Observations of Pluto were obtained on four nights in June 1974 and twelve nights in January through March 1975 with the University of Iowa 60-cm reflector. An average of the nightly infrared reflectance values, weighed by their standard deviations, gave the ratio of the reflectance at 0.86 µm relative to that at 0.55 µm as 1.33±0.04. This result is tabulated with the reflectance measurements of Hardie in Table 20 and plotted in Fig. 6.4.

Figures 6.4 and 6.5 show the reflectance of Pluto compared to several asteroids and meteorites of various reflectance types. These comparisons indicate that there are objects in the solar system whose reflectances resemble that of Pluto between 0.3 µm and 0.86 µm. The best match is with the asteroid 5 Astrae, an iron meteorite with low nickel content, and a stony iron meteorite. The correlation between Pluto and the iron meteorites is interesting in view of Manning's (1971) earlier observation of the general and specific resemblance of the reflectance spectrum of Pluto to the absorption spectrum of a terrestrial iron-bearing silicate.

Noting the similarity of Pluto and Triton, in terms of their size, mass, brightness, distance from the Sun, and equilibrium temperature, Benner, Fink, and Cromwell (1978) suggested that

TABLE 20. Pluto Relative Reflectance Measurements of Hardie[8]

Filter	λ_{eff} (μm)	Full Width (μm)	Relative Reflectance (in mag.)
U	0.37	0.07	$+0^m.30$
B	0.44	0.09	$+0.17$
V	0.55	0.07	0.00
R	0.68	0.23	-0.18
I_H	0.82	0.16	-0.17
I_L	0.86	0.04	-0.31

[8] Lane, W.A., Neff, J.S., and Fix, J.D. A measurement of the relative reflectance of Pluto at 0.86 micron. *Publ. Astron. Soc. Pac.*, 88, 77 (1976). By permission of the authors and the editor of *Publications*.

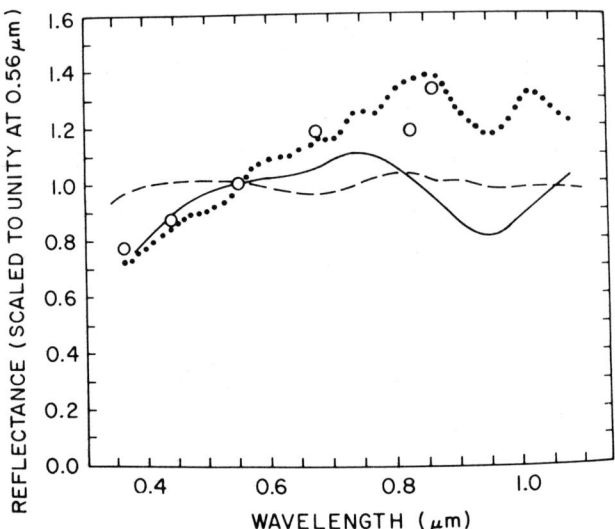

Fig. 6.4. Comparison of the reflectance of Pluto (open circles), including the *UBVRI* measurements of Hardie and the result of this study, to the following asteroids: dashed line = 2 Pallas, solid line = 4 Vesta, and dotted line = 5 Astrae. We have drawn smooth curves for asteroid data given by Chapman, McCord, and Johnson (1973).[9]

---

[9]Lane, W.A., Neff, J.S., and Fix, J.D. A measurement of the relative reflectance of Pluto at 0.86 micron. *Publ. Astron. Soc. Pac.*, 88, 77 (1976). By permission of the authors and the editor of *Publications*.

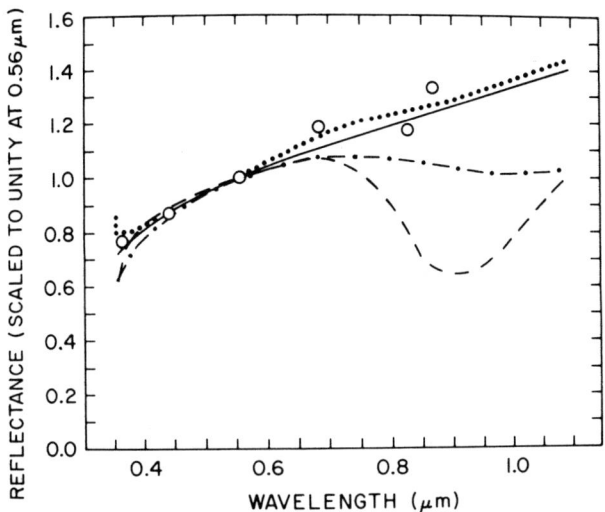

Fig. 6.5. Comparison of the reflectance of Pluto (open circles) to the following meteorites: dotted line = stony iron (Veramin), solid line = iron (Casey County), dot-dashed line = type C30 carbonaceous chrondrite (Kainsaz), and dashed line = type 2 eucrite (Bereba).[10]

---

[10] Lane, W.A., Neff, J.S., and Fix, J.D. A measurement of the relative reflectance of Pluto at 0.86 micron. *Publ. Astron. Soc. Pac.*, 88, 77 (1976). By permission of the authors and the editor of *Publications*.

any possible atmosphere on either object could therefore be expected to have similar characteristics. Benner and his colleagues considered a methane atmosphere to be the most likely possibility, particularly in view of Cruikshank, Pilcher, and Morrison's (1976) evidence for methane frost on the surface of Pluto, implying a minimum methane atmosphere in equilibrium with this frost. They decided therefore to search for such an atmosphere by obtaining new infrared spectra of Pluto and Triton for comparison with laboratory spectra of methane.

Observations of Pluto, Triton, and four solar-type comparison stars were carried out on June 18 and 19, 1976 with the 229-cm telescope of Steward Observatory on Kitt Peak. An 800 lines/mm diffraction grating spectrograph with a dispersion of ~100 Å/mm was used in conjunction with a cooled, three-stage Varo image tube and the spectra were recorded on Eastman Kodak IIa-D spectroscopic plates. The spectral resolution of the combined system was ~5 Å and the 25 mm field provided spectral coverage from 6800 to 9200 Å. Neon and argon comparison spectra as well as night sky lines were used for wavelength calibration. The night sky emission proved unexpectedly strong and had to be eliminated from the desired spectra in two steps. First, a small aperture (2x3 arcsec) was used at the entrance slit, admitting not much more than the object image. Then a night sky spectrum was recorded simultaneously and later used for a point by point subtraction of the night sky component from the object spectra. Because all the object spectra had been positioned identically on the image tube, it was possible to average the spectra and perform ratios without resulting in distortions which might arise from the nonuniform image tube sensitivity across its face. The average of the two Pluto exposures, an average of the stellar comparison spectra, and their ratio are shown in Fig. 6.6.

The apparent feature at 8410 Å in the Pluto spectrum is due to a plate defect. The additional feature on the shortwave side of 8410 Å and near 8540 Å are too close to solar, telluric, or night sky features to permit a definite conclusion. Benner and his co-workers found the general appearance of the ratio spectra of Pluto and Triton rather similar, both rising somewhat between 6800 and 7400 Å before leveling off and decreasing toward 9000 Å. The reddening trend agreed quite well with *UBVRI* photometry summarized by Harris (1961). Their results did not support the measurements of Lane, Neff, and Fix (1976) who claimed a substantial increase in albedo at 8600 Å. Pos-

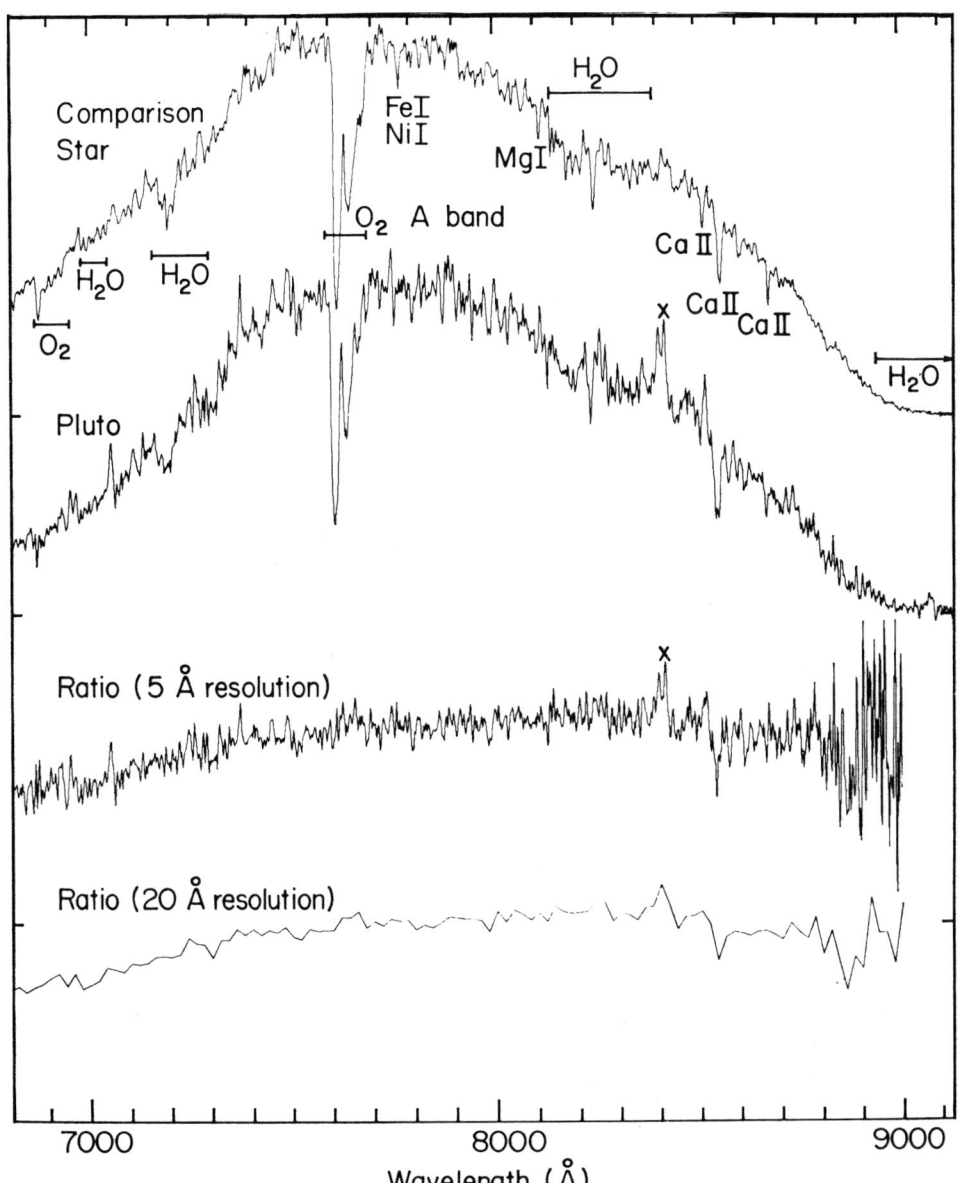

Fig. 6.6. Spectra of Pluto, an average of four solar type comparison stars (top), and their ratios at the observed resolution of 5 Å and a smoothed resolution of 20 Å. Telluric and solar absorptions are marked on the comparison star. The zero levels are displaced as indicated on the side. A plate defect is indicated by X.[11]

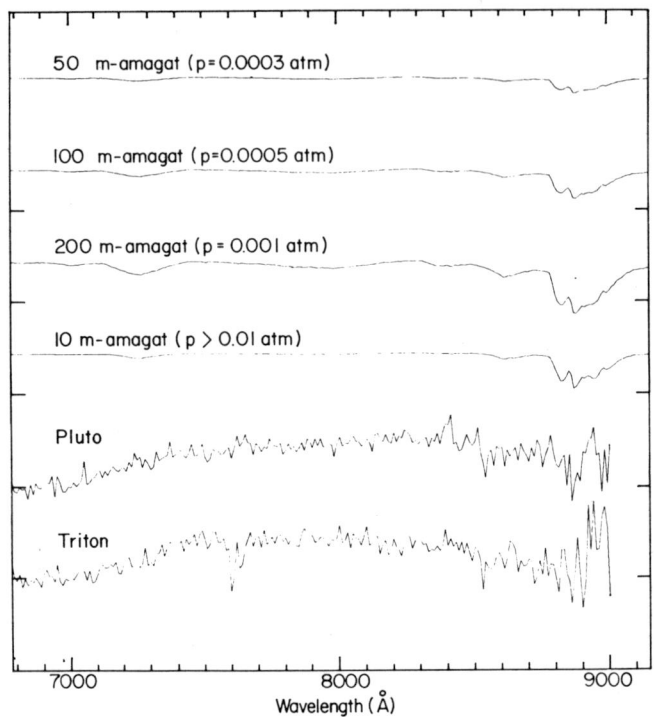

Fig. 6.7. Comparison of synthetic methane spectrum calculations and the ratio spectra of Pluto and Triton all at a resolution of 10 Å. The top three synthetic spectra assume a pressure produced by the methane atmosphere alone. The fourth one assumes a pressure >0.01 atm so that pressure effects can be neglected.[12]

[11,12] Benner, D.C., Fink, U., and Cromwell, R.H. Image tube spectra of Pluto and Triton from 6800 to 9000 Å. *Icarus*, 36, 82 (1978). By permission of the authors and *Icarus*. (Copyright [1978] by Academic Press, Inc.).

sibly the strongest, and definitely the most important, feature occurs at 8870 Å in the Pluto spectrum and with somewhat less certainty in Triton's spectrum. It coincides with the strongest laboratory feature in the methane spectrum up to 1 μm (Dick and Fink, 1977; Fink, Benner, and Dick, 1977). It was the original intention of Benner and his colleagues' (1978) investigation to obtain good spectral coverage of this feature at 8870 Å, since it would provide the best test, below 1 μm, for a methane atmosphere. Unfortunately, the steeply diminishing response of the image tube at that wavelength casts some doubt on the reality of this feature.

Since the effect of atmospheric pressure on any abundance determination can be significant, Benner and his colleagues analysed their data first by assuming a substantial pressure and then considering a much lower one. One of the principal results of their laboratory spectral analysis of methane (Dick and Fink, 1977; Fink, Benner, and Dick, 1977) was the absence of a strong pressure effect on the absorption features of methane. From new calculations of synthetic spectra they found that above 0.01 atm the absorptions were unaffected by the pressure, while below this pressure its effect noticeably weakened the absorptions. The effect of the pressure can therefore be neglected for the analysis of methane absorptions where little contribution to the absorption occurs below 0.01 atm. Total atmospheric pressures of that order could be produced on Pluto by the presence of infrared-inactive broadening gases such as neon. Figure 6.7 shows a best-fit synthetic spectrum for Pluto and Triton at a pressure of 0.02 atm and an equivalent methane absorption of 10 m-amagat. It can be seen that a much higher abundance is required to match the absorption on Pluto or Triton if the pressure effect is included. Examination of the spectra shows that an abundance between 100 and 200 m-amagat, assuming a pressure produced by the methane atmosphere only, matches the possible absorption on Pluto quite well.

Under the assumption of the reality of the 8900 Å absorption feature, Benner and his colleagues considered the nature and implications of a methane atmosphere on Pluto. They concluded that it was quite possible for a methane atmosphere to be produced by the sublimation of a surface frost of methane. Because of Pluto's high eccentricity and, hence, varying distance from the Sun, a substantial atmosphere might only exist near Pluto's perihelion. They also examined the possibility that the absorption on Pluto was caused by a layer of methane frost rather than a methane atmosphere. Their laboratory investigations of solid and gaseous methane absorptions from 1 to 4 µm showed that the gaseous and solid absorption features occurred at the same wavelengths. Therefore, the possibility exists that the absorptions detected on Pluto by Cruikshank, Pilcher, and Morrison (1976) are indicative of either a methane atmosphere, a surface frost of methane, or a combination of the two. Benner and his co-workers suggested that until the gaseous and solid absorption features in the 1 to 3 µm region can be distinguished, the methane absorption of 8900 Å offered the best possibility of deciding between the two; certainly, they said, further spectral studies, with new instruments now being developed, should be undertaken. They observed that, should the feature at 8900 Å turn out to be weaker, or even nonexistent, their analysis would be applicable, instead to an upper limit for methane on Pluto.

Nacozy and Diehl (1978) presented a semianalytical solution for the long-term motion of Pluto. They introduced a new model consisting of a modified circular restricted three-body problem having a hypothetical oblate Sun and Neptune as primaries, with Neptune in a circular orbit on the same plane as the solar equator, and Pluto with negligible mass. This was the simplest dynamical model they could define that gave the long-term motion of Pluto, neglecting short-period terms and very long-period terms. Several solutions can be derived from the model, one which includes both the Pluto-Neptune and Pluto-Uranus resonant effects, as well as the nonresonant effects due to Jupiter and Saturn. This solution gives the long-term libratory motion of the eccentricity, inclination, and perihelion of Pluto. The eccentricity librates between a maximum of 0.266 and a minimum of 0.218. The maximum and minimum values for the inclination are $17°.15$ and $14°.67$, respectively, with a libration centre at $16°.03$. Unlike the libration centre in eccentricity the libration centre in inclination occurs at a value greater than the average inclination.

Nacozy and Diehl found that the long-term librational character of Pluto's elements would not be altered by the introduction of small gravitational forces (such as changes in the eccentricity and inclination of Neptune), or very small nongravitational, dissipative forces (see Whipple, 1950; Kiang, 1973), or adjustments of initial conditions, masses, and other parameters. Nor is the avoidance of close encounters between Pluto and Neptune affected. So the long-term stability of Pluto's motion was confirmed.

On July 7, 1978 *Circular 3241* of the *International Astronomical Union's Central Bureau for Astronomical Telegrams* announced the discovery of a satellite to Pluto by James W. Christy of the U.S. Naval Observatory. This was one of the most important and long-awaited discoveries in solar-system astronomy this century. For, at long last, here was a way to calculate the mass of Pluto with far greater precision than from perturbation theory or diameter/density estimates. Indeed, less than two weeks after the IAU announcement, D.W. Hughes (1978) provided just such a calculation by the classical equation (1):

$$M_p = 4\pi^2 d^3/GT^2, \qquad (1)$$

where d is taken to be the mean satellite-planet distance, T is the satellite's orbital period, G is Newton's gravitational constant and the satellite is assumed to be much less massive than the planet. Substituting the values published in *IAU Circular 3241*, Hughes computed Pluto's mass to be 1.56×10^{22}g, a value consistent with the uncertainty limits of the mass values given by Cruikshank, Pilcher and Morrison (1976). Hughes suggested the name of Persephone, Pluto's queen in Hades, for the new satellite.

Christy (Christy and Harrington, 1978) had noticed elongations of the photographic images of Pluto on plates taken with the U.S. Naval Observatory's 1.55-m astrometric reflector at Flagstaff, Arizona during a program of regular astrometric observations of Pluto begun in 1978. While measuring these plates on an automatic measuring machine on June 22, Christy noticed a consistent elongation taking the form of a faint southerly extension on all exposures made on April 13 and 20 and a faint northerly extension on all exposures of May 12. The possibility of a satellite was suggested after guiding errors were ruled out by the roundness of the star images and the possibility of a faint background star was eliminated by a check of the Palomar Sky Survey prints. Verification of the elongation was then sought on plates of Pluto taken previously with the 1.55-m telescope in 1965, 1970, and 1971 for other projects. Because the elongation in 1978 was only of the order of 0".7 on the best images, it was apparent that any elongation would be detected only under optimal seeing conditions. Nevertheless Christy and Harrington were able to detect and measure the direction of elongation on two plates taken in 1965 and five plates taken in 1970. Examination of the 1970 plates suggested a period of just over six days.

Following these initial observations, Pluto was observed with the 1.55-m instrument every possible night by C.C. Dahn although it was well past the meridian at the end of evening twilight. Definite elongations were observed on two nights and another estimate of the extent of the elongation was made on July 2. Estimates by various observers ranged from 0".8 to 0".9 with the generally agreed best estimate being the lower one.

Figure 6.8 shows the best image obtained that night.

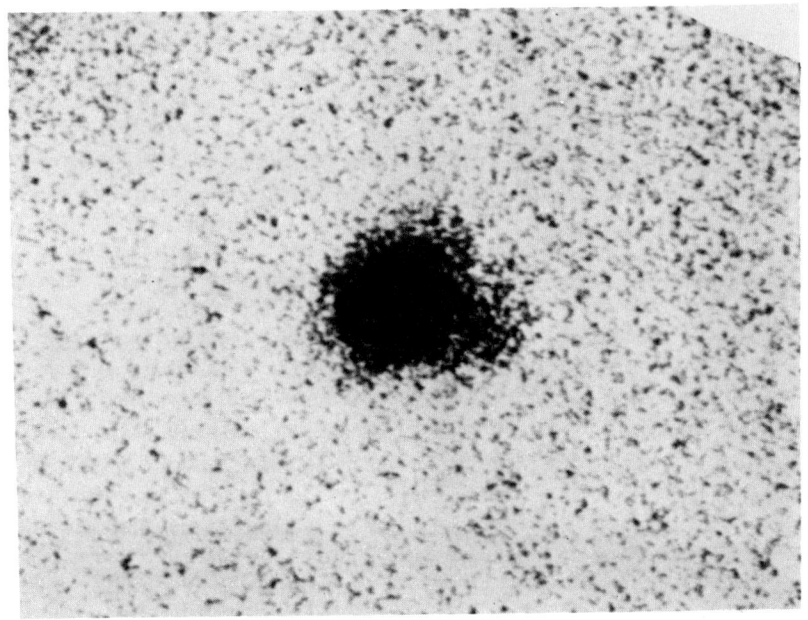

Fig. 6.8. Image of Pluto from plate taken 2 July 1978 with USNO 1.55-m astrometric reflector; magnification is approximately 100 times. Satellite appears as flare on planet image in about 4:30 position.[13]

---

[13]Christy, J.W., and Harrington, R.S. The satellite of Pluto. *Astron. J.*, 83, 1005 (1978). By permission of the authors and *Astronomical Journal*.

An elongation of 0".8 corresponds to a distance of 17,000 km at the distance of Pluto, three times the upper limit for the planet's diameter established earlier (Halliday, Hardie, Franz, and Priser, 1966), neatly eliminating the possibility of the extension being an effect of Pluto's rotation or variable surface features. J.A. Graham (1978) reported detecting an elongated image with a mean diameter of 1".6, an axis ratio slightly less than 2:1, and a position angle of approximately 166°±5° on plates taken at the prime focus of the Cerro-Tololo 4-m reflector on July 6. D.J. Mulholland, at McDonald Observatory, likewise photographed Pluto on July 6 and confirmed the reality of the image phenomenon. Subsequently, on the same day, the discovery was announced to the astronomical community and the media by Captain J.C. Smith, USN.

In an attempt at an initial interpretation, Christy and Harrington assumed that the 6.3867-day period associated with the light curve (Andersson and Fix, 1973) also applied to the satellite, that the orbit was circular with a radius equal to the maximum observed separation, that the time and direction of maximum elongation correspond to nodal passage, and that the inclination is high enough to permit only limited directions of elongation but not so high as to affect the light curve. From all these considerations the following elements were adopted (Table 21).

TABLE 21. Elements of Pluto's Satellite

a = 0".8 at a distance of 30 a.u.

P = 6.3867 days

e = 0.0

i = ±105°
} with respect to the plane of
Ω = 350°   the sky at α = 13:30, δ = +10°.5.

The inclination to the ecliptic is approximately 115° or 55°, with the pole of the orbit lying near either α = 8^H, δ = -5° or

$\alpha = 19^H$, $\delta = 35°$, both in the domain of the pole of rotation of Pluto suggested by Andersson and Fix (1973). The first orientation predicts transits and occultations in the years 1983-1987, the second in the years 1968-1972. Using these elements, Christy and Harrington computed an ephemeris for the times of all positive observations (those in which elongations were detected), as well as for the time of Kuiper's (1950) attempt to measure Pluto's diameter with the 5-m Palomar reflector. The predicted position angles and separations, along with those observed, if any, are given in Table 22.

TABLE 22. Predicted and Observed Positions of Pluto Satellite[14]

Date		P_{calc}	P_{obs}	D_{calc}	D_{obs}
1950 Mar	22.20	120°		0".21	
1965 Apr	29.16	167	160°	0.72	
1965 May	1.15	8	10	0.46	
1970 Jun	13.17	171	160	0.76	
1970 Jun	15.17	23	20	0.31	
1970 Jun	16.17	353	350	0.74	
1970 Jun	17.16	338	340	0.58	
1970 Jun	19.16	176	170	0.69	
1978 Apr	13.31	172	170	0.79	
1978 Apr	20.29	164	160	0.73	
1978 May	12.22	350	350	0.80	0".7
1978 Jul	1.20	14	10		
1978 Jul	2.17	352	350	0.80	0.8
1978 Jul	5.15	175	175	0.76	
1978 Jul	6.00	162	166	0.71	

[14] Christy, J.W., and Harrington, R.S. The satellite of Pluto. *Astron. J.*, 83, 1005 (1978). By permission of the authors and *Astronomical Journal*.

The agreement between prediction and observation is, as Christy and Harrington say, remarkable considering the difficulty of the measurement.

From the period and mean distance given in Table 21, Christy and Harrington calculated a mass for Pluto plus satellite of approximately 200,000,000 reciprocal solar masses (about 0.0017 earth masses). If the diameter of Pluto is taken as 3000 km (Cruikshank, Pilcher, and Morrison, 1976), then the mean density of the planet is about 0.7 times that of water. This would suggest that Pluto, like the satellites of the outer major planets, consists predominantly of frozen volatiles. If it is also assumed that Pluto's light variation is indeed due to the planet's rotation (the satellite could be no more than 1.6 magnitudes fainter in the visual to produce a variation of 0.22 in the total V), then the satellite revolution and the planet rotation are synchronous. Such a condition would be inherently stable. Christy and Harrington calculated that, with a magnitude difference somewhat greater than 2, and assumed albedo and density similar to Pluto's, the diameter of the satellite would be around 0.4 that of Pluto and the mass would be 0.05-0.10 that of Pluto. They claim that the satellite-like nature of Pluto lends support to the idea that Pluto may be an escaped satellite of Neptune.

Christy proposed that the satellite be named *Charon* after the ferryman who, according to classical Greek mythology, carried the souls of dead people in his boat across the river Styx to Hades, the kingdom of Pluto.

Lawton (1978) has provided a detailed description of the Pluto-Charon system. He notes that Charon is even more massive relative to Pluto than our moon is relative to the Earth (1:81.4). Charon may, if its revolution period is indeed equal to Pluto's period of rotation, be unique in that it is the only known synchronous natural satellite in the solar system. Lawton notes that Charon would be a most impressive sight from the surface of Pluto, appearing five times the diameter of the moon in our skies with a visual magnitude of between -9 and -9.5 - about as bright as our moon at first quarter phase. Charon would also appear to go through phases similar to our moon, with a cycle of 6.39 days. Lawton also points out that Pluto is not the "gloomy dark world" it is often depicted to be. The Sun appears as a brilliant star with a *mean* visual magnitude of about -19, about 1600 times the luminosity of the full moon on Earth.

Following the discovery of Charon, calculations by Harrington and Van Flandern (1978) suggested that Pluto and its satellite could have been created by an encounter of Neptune with a hypothetical 10th planet of 3 to 4 earth masses. Such a planet passing by the Neptunian system could have perturbed Triton into its present retrograde orbit and ejected another satellite (Pluto) into an orbit that could have subsequently locked into its present resonant orbit. Tidal stresses may also have broken off a portion of Pluto that then became its satellite. A further consequence would be an irregularly-shaped planet, with a brightness variation caused by its rotation. Lawton (1979) noted that if the Harrington-Van Flandern encounter occurred shortly after the formation of the solar system then there has been adequate time for the hypothetical planet to have reached equilibrium. Lawton says the simplest way such a body could be stabilized would be for it to complete a single orbit commensurate with Pluto and Neptune in the ratio of 1:2 and 1:3, respectively. He points out that although the hypothesis of Harrington and Van Flandern is open to criticism, it does offer the possibility of the new planet having an extremely eccentric orbit and therefore permit the reconciliation of Lalande's two 1795 observations of Neptune. The simplest - and most naive - assumption would be that the planet was close to Neptune during 1795 and caused the perturbation. Lawton proposed a highly speculative set of elements for "Planet X". He conceived a planet of 5 to 6 earth masses with a diameter of 15,000 to 18,000 km and a magnitude of 11 at perihelion. The high eccentricity ($e = 0.35$) of Planet X's 495-year orbit would be reflected in its perihelion (30 a.u.) and aphelion (92.5 a.u.).

Chapter 7
Full Circle
1980-2178

On January 23, 1979 Pluto's highly eccentric orbit crossed that of Neptune's so that both planets were equidistant from the Sun, at 30.3 a.u. Pluto will reach perihelion (29.6 a.u.) in September, 1989. The two planets will again be equidistant on March 15, 1999 after which Neptune will relinquish its temporary role as the outermost known planet in the solar system.

The Department of Astronomy at New Mexico State University, Las Cruces, New Mexico, has planned a special symposium on Pluto in February, 1980 to commemorate the 50th Anniversary of the Discovery of Pluto. A list of the papers to be presented was unavailable at the time of writing (July, 1979).

The first observations of Pluto allowing direct measurement of its diameter and resolution of Charon are likely to be made with the Space Telescope (ST) scheduled to be launched by the Space Shuttle in 1983. The resulting refinement of Charon's elements should permit the calculation of Pluto's mass with accuracy comparable to that obtained by the same method for the other outer planets. However the resolution of surface details on Pluto and Charon will have to await future unmanned spacecraft of the Voyager or Pioneer class. Unfortunately this remains a distant prospect as no Pluto missions are currently planned or under consideration. A "Grand Tour" space mission designed to take advantage of an exceptionally favorable alignment of the outer planets, which occurs only once about every 179 years, had received serious consideration during the early 1970's. A typical Grand Tour mission would have involved targeting spacecraft past Jupiter on toward several possible combinations of Saturn, Uranus, Neptune, and Pluto using the gravitational acceleration of each planetary encounter to boost the spacecraft on toward the next planet. Originally, sophisticated spacecraft called TOPS (Thermoelectric Outer Planet Spacecraft), having self-monitoring and self-repair capabilities, would have been launched in 1977 to

fly past Jupiter, Saturn, and Pluto. This would have cut the
flight time to Pluto from a prohibitive 41 years by the most
economical trajectory to a mere 9 years. Had such a mission
been launched as scheduled we could now be anxiously antici-
pating the arrival of a spacecraft at the Pluto-Charon system
in 1986. Unfortunately, due to budgetary constraints, the
National Aeronautics and Space Administration (NASA) was unable to plan a Grand Tour mission as originally envisaged and
opted instead for a scaled-down version. This new mission
consisting of a pair of Mariner class spacecraft, Voyagers 1
and 2, has been spectacularly successful at the time of
writing, returning a wealth of scientific data on Jupiter and
its satellites. Both spacecraft have continued onward to
Saturn utilizing the gravitational acceleration provided by
Jupiter to reduce the flight time to Saturn to only two years.
Depending upon the success of Voyager 1's reconnaissance of
Saturn, Voyager 2 has the option of being targeted towards a
flypast of Uranus in January, 1986 with the possibility of
being redirected yet again to an encounter with Neptune in
September, 1989.

The first spacecraft to cross Pluto's orbit, in 1987, will be
Pioneer 10 which left the earth in 1972. Pioneer 10 will then
leave the presently known boundary of our solar system with a
residual velocity of 11.5 km/sec becoming, in effect, our
first interstellar spacecraft. However, neither Pioneer 10,
nor its followers Pioneer 11 and Voyagers 1 and 2, will pass
close enough to Pluto to provide any useful data except on the
nature of the interplanetary medium through which the planet
travels. Provided that their power supplies continue to de-
cline at no more than the projected rate and their large antennae maintain their orientation, these spacecraft will be
trackable out to the distance of Pluto with the Deep Space
Network (DSN) of tracking stations and to even further distances with the 300-m Arecibo antenna or the Very Large Array
(VLA) of radio telescopes in New Mexico.

Although the opportunity afforded by a unique alignment of the
outer planets to fly a very short flight time mission to Pluto
has passed, the opportunity of flying a single planet gravity
-assist mission of the Pioneer 10 type reoccurs about every
twelve years, the revolution period of Jupiter. Such a Pluto
mission might utilize updated Voyager or Jupiter Orbit and
Probe (JOP) class spacecraft and would have a flight time of
about 15 years although this parameter will increase after
Pluto passes perihelion in 1989 and begins to recede into

deeper space. It is very probable that a Pluto orbital mission, rather than a simple flypast trajectory, will be chosen because such a mission would provide very accurate information on the mass, size and shape of Pluto and Charon as well as detailed imagery of their surfaces. Because the formation of the Pluto-Charon system was likely a violent event possibly involving escape from the Neptune system, Charon may be an irregularly-shaped piece of planetary debris similar to the Martian satellites Deimos and Phobos. Pluto itself will likely have experienced heavy meteoroid bombardment during its formation or subsequent history particularly since this appears to be a consistent feature of solar system history, at least for the inner planets and satellites and Jupiter's Galilean satellites. Pluto may also exhibit the hemispherical asymmetry of the inner planets and our moon in which one hemisphere appears radically different geologically from the other. Such a hemisperical difference on Pluto might account for its 6.39-day light variation as the planet rotates. An orbital mission would allow ample radio-occultation experiments to determine the nature and composition of the Plutonian atmosphere if present. A flypast mission would seem particularly wasteful in view of the lengthy mission flight time and the energy requirements for orbital insertion around Pluto may not be excessive owing to the fairly low residual velocity of the spacecraft upon arrival at Pluto.

A Pluto space mission, despite its technological feasibility, remains a remote prospect at time of writing since no such mission is currently planned or under consideration by NASA. However a launch before the end of the 1980's could still provide us with detailed knowledge of the Pluto-Charon system by the end of this century. In 2178 A.D., Pluto will have come full circle since its discovery in 1930.

Bibliography

Adams, A.N. & Scott, D.K. *U.S. Naval Observ. Circ.*, Nos. 103, 105, 115, 118 (1964-1968).
Airy, G.B. Report on astronomy. *Brit. Assoc. Advanc. Sci. Rep.*, 1832 (1832).
Airy, G.B. Account of some circumstances historically connected with the discovery of the planet exterior to Uranus. *Mon. Not. R. Astron. Soc.*, 7, 121 (1846).
Alter, D. The story of Pluto. *J. R. Astron. Soc. Canada*, 46, 1 (1952).
Andersson, L.E. The orientation of the rotational axis of Pluto. *Bull. Am. Astron. Soc.*, 5, 36 (1973).
Andersson, L.E. Eclipse phenomena of Pluto and its satellite. *Bull. Am. Astron. Soc.*, 10, 586 (1978).
Andersson, L.E. & Fix, J.D. Pluto: New photometry and a determination of the axis of rotation. *Icarus*, 20, 279 (1973).
Ash, M.E.; Shapiro, I.I. & Smith, W.B. The system of planetary masses. *Science*, 174, 551 (1971).
Baade, W. Beobachtungen des Pluto. *Astron. Nachr.*, 242, 367 (1930).
Baade, W. The photographic magnitude and color index of Pluto. *Publ. Astron. Soc. Pac.*, 46, 218 (1934).
Bakos, G. Observations of asteroids and Pluto made with an image orthicon tube. *Astron. J.*, 70, 171 (1965).
Baldet, M. *J. Brit. Astron. Assoc.*, 40, 248 (1930).
Baldi, P. & Caputo, M. The hydrostatic equilibrium and its implications on the mechanical properties of planets. *Ann. Geofis.*, 27, 235 (1974).
Banachiewicz, T. Orbit of Pluto. *C. R. Acad. Sci. (Paris)*, 191, 246 (1930a).
Banachiewicz, T. Trans-Neptunian planet Pluto. *C. R. Acad. Sci. (Paris)*, 191, 248 (1930b).
Banachiewicz, T. Orbit of the trans-Neptunian planet. *C. R. Acad. Sci. (Paris)*, 191, 319 (1930c).
Banachiewicz, T. Photographic observations of Pluto. *Acad. Polonaise Sci. et Lettres, Bull.*, 1-2A, 23 (1936).

Barbieri, C.; Capaccioli, M.; Ganz, R. & Pinto, G. Astrometric programs being carried out at the Padova and Asiago Observatories. *Mem. Soc. Astron. Ital.*, 43, 635 (1972a).

Barbieri, C.; Capaccioli, M.; Ganz, R. & Pinto, G. Accurate positions of the planet Pluto in the years 1969-1970. *Astron. J.*, 77, 521 (1972b).

Barbieri, C.; Capaccioli, M. & Pinto, G. Astrometric positions of the planet Pluto in the years 1971-74. *Astron. J.*, 80, 412 (1975).

Belot, E. Origin and formation of Pluto according to dualistic cosmogony. *C. R. Acad. Sci. (Paris)*, 191, 248 (1930).

Bessel, F.W. *Fundamenta astronomiae pro A. 1755 deducta ex observationibus viri incomparibilis James Bradley in specula astronomica Grenoricensi per A. 1750-62 institutis*, Regiomonti, 1818.

Biesbroeck, G. van. The mass of Neptune from a new orbit of its second satellite Nereid. *Astron. J.*, 62, 272 (1957).

Biesbroeck, G. van. Positions of planet Pluto. *Astron. J.*, 68, 738 (1963).

Biesbroeck, G. van; Vesely, C.D.; Marsden, B.G. & Aksnes, K. Observations of comets, minor planets, Pluto, and satellites. *Astron. J.*, 81, 122 (1976).

Bouvard, A. *Tables astronomiques publiees par le Bureau des Longitudes de France contenant les Tables de Jupiter, de Saturne et d'Uranus, construites d'apres la theorie de la Mecanique celeste.* Paris, 1821.

Bower, E.C. On the orbit and mass of Pluto with an ephemeris for 1931-1932. *Lick Obs. Bull.*, 15, 171 (1931).

Bower, E.C. Ephemeris of Pluto for 1932-33 - fourth paper. *Lick Obs. Bull.*, 16, 31 (1932).

Bower, E.C. Pluto - ephemeris for 1933-34 - fifth paper. *Lick Obs. Bull.*, 16, 115 (1933).

Bower, E.C. Pluto - emphemeris for 1934-35 - sixth paper. *Lick Obs. Bull.*, 17, 53 (1934).

Bower, E.C. & Whipple, F.L. Preliminary elements and ephemeris of the Lowell Observatory object. *Lick Obs. Bull.*, 14, 189 (1930a).

Bower, E.C. & Whipple, F.L. *Publ. Astron. Soc. Pac.*, 42, 239 (1930b).

Bower, E.C. & Whipple, F.L. Elements and ephemeris of the Lowell Observatory object (Pluto) - second paper. *Lick Obs. Bull.*, 15, 35 (1931).

Brady, J.L. The effect of a trans-Plutonian planet on Halley's comet. *Publ. Astron. Soc. Pac.*, 84, 314 (1972).

Brady, J.L. & Carpenter, E. The orbit of Halley's comet and the apparition of 1986. *Astron. J.*, 76, 728 (1971).

Brouwer, D. The motions of the outer planets. *Mon. Not. R. Astron. Soc.*, 115, 221 (1955).
Brouwer, D. In: *The Theory of Orbits in the Solar System and in Stellar Systems*, G. Contopoulos, Ed., p. 227. Academic Press, New York, 1966.
Brouwer, D. & Clemence, G.M. In: *Planets and Satellites*, G.P. Kuiper and B.M. Middlehurst, Eds., p. 63. Univ. of Chicago Press, Chicago, 1961.
Brown, E.W. Predictions of trans-Neptunian planets from perturbations of Uranus. *Nat. Acad. Sci. Proc.*, 16, 364 (1930).
Brown, E.W. Criterion for the prediction of an unknown planet. *Mon. Not. R. Astron. Soc.*, 92, 80 (1931).
Cacciatore, N. Letter. *C. R. Acad. Sci. (Paris)*, 2, 154 (1836).
Cameron, A.G.W. The formation of the sun and planets. *Icarus*, 1, 13 (1962).
Challis, J. Letter to the Editor. *Athenaeum*, 1300 (1846).
Christy, J.W. & Harrington, R.S. The satellite of Pluto. *Astron. J.*, 83, 1005 (1978).
Cochran, W.D. & Light-Cochran, A. Digicon spectroscopy of Triton and Pluto. *Bull. Am. Astron. Soc.*, 10, 585 (1978).
Cohen, C.J. & Hubbard, E.C. Libration of Pluto-Neptune. *Science*, 145, 1302 (1964).
Cohen, C.J. & Hubbard, E.C. Libration of the close approaches of Pluto to Neptune. *Astron. J.*, 70, 10 (1965).
Cohen, C.J.; Hubbard, E.C. & Oesterwinter, C. New orbit for Pluto and analysis of differential corrections. *Astron. J.*, 72, 973 (1967).
Cohen, C.J.; Hubbard, E.C. & Oesterwinter, C. On the stability of Pluto's orbit. *(Private communication)*, 1979.
Colgrove, W.G. Letter to the Editor. *J.R. Astron. Soc. Canada*, 31, 55 (1937).
Colombo, G. & Franklin, F.A. *Periodic orbits, stability and resonances, Sao Paulo, Brazil, 4-12th. September, 1969.* D. Reidel, Dordrecht, Netherlands, 1970.
Crommelin, A.C. Pluto. *J. Brit. Astron. Assoc.*, 41, 116 (1931a).
Crommelin, A.C. Examination of the perturbations produced by Pluto on Saturn and Jupiter. *J. Brit. Astron. Assoc.*, 41, 221 (1931b).
Crommelin, A.C. The discovery of Pluto. *Mon. Not. R. Astron. Soc.*, 91, 380 (1931c).
Cruikshank, D.; Pilcher, C.B. & Morrison, D. Pluto: Evidence for methane frost. *Science*, 194, 835 (1976).

Dauvillier, A. The nature of Pluto and Triton. *C. R. Acad. Sci. (Paris)*, 233, 901 (1951).

Dermott, S.F. Pluto, Herculina, Mercury and Venus: Their real and imaginary satellites. *Bull. Am. Astron. Soc.*, 10, 586 (1978).

Dollfus, A. In: *Planets and Satellites*, G.P. Kuiper and B.M. Middlehurst, Eds., p. 343. Univ. of Chicago Press, Chicago, 1961.

Duncombe, R.L.; Klepczynski, W.J. & Seidelmann, P.K. Orbit of Neptune and the mass of Pluto. *Astron. J.*, 73, 830 (1968a).

Duncombe, R.L.; Klepczynski, W.J. & Seidelmann, P.K. Mass of Pluto. *Science*, 162, 800 (1968b).

Duncombe, R.L.; Klepczynski, W.J. & Seidelmann, P.K. Note on the mass of Pluto. *Publ. Astron. Soc. Pac.*, 82, 916 (1970).

Duncombe, R.L.; Klepczynski, W.J. & Seidelmann, P.K. A determination of the masses of the five outer planets. *Celestial Mech.*, 4, 224 (1971).

Eckert, W.J.; Brouwer, D. & Clemence, G.M. *Astron. Papers Am. Ephem. Naut. Almanac*, 12 (1951).

Eichelberger, W.S. & Newton, A. *Astron. Papers Am. Ephem. Naut. Almanac*, 9 (1926).

Emanuelli, P. The cometary aphelia and the trans-Neptunian planet. *Pop. Astron.*, 38, 451 (1930).

Esclangon, E. Trans-Neptunian planet. *C. R. Acad. Sci. (Paris)*, 190, 834, 897 & 957 (1930a).

Esclangon, E. Determination of position and elements of a planet or distant star. Application to the Lowell celestial body. *C. R. Acad. Sci. (Paris)*, 190, 981 (1930b).

Esclangon, E. Determination of the position and elements of a planet or comet by three corresponding observations of a small arc of the orbit. *C. R. Acad. Sci. (Paris)*, 190, 1085 (1930c).

Esclangon, E. Orbit of the trans-Neptunian planet. *C. R. Acad. Sci. (Paris)*, 191, 629 (1930d).

Fayet, G. Orbits of comets and the planet Pluto. *C. R. Acad. Sci. (Paris)*, 192, 1362 (1931a).

Fayet, G. Orbits of the planets Neptune and Pluto. *C. R. Acad. Sci. (Paris)*, 193 (1931b).

Fix, J.D. Comments on the interior of Pluto. *Icarus*, 16, 569 (1972).

Fix, J.D.; Neff, J.S. & Kelsey, L.A. Spectroscopy of Pluto. *Astron. J.*, 75, 895 (1970).

Fixlmillner, P. *Berliner Astron. Jahrb.*, 249 (1787).

Flammarion, C. *L'Astronomie*, 3, 81 (1884).

Forbes, G. On comets. *Proc. R. Soc. Edinburgh*, 10, 426 (1880a).

Forbes, G. Additional note on an ultra-Neptunian planet. *Proc. R. Soc. Edinburgh*, 11, 89 (1880b).
Foss, A.; Shaw-Taylor, L. & Whitworth, D. Search for a trans-Plutonian planet. *Nature*, 239, 266 (1972).
Gaillot, M.A. Contribution a la recherche des planetes ultra-neptuniennes. *C. R. Acad. Sci. (Paris)*, 148, 754 (1909).
Gaillot, M.A. *Mem. Obs. Paris*, 28, 83 (1910).
Goldreich, P. & Soter, S. Q in the solar system. *Icarus*, 5, 375 (1966).
Golitsyn, G.S. A possible atmosphere on Pluto. *Sov. Astron. Lett.*, 1, 19 (1975).
Graff, K. Helligkeit des Pluto. *Astron. Nachr.* 240, 163 (1930).
Graham, J.A. In: *Circular 3241, International Astronomical Union Central Bureau for Astronomical Telegrams.* July 7, 1978.
Gunn, E.J. Another planet? *New Sci.*, 48, 345 (1970).
Halliday, I. A proposal for a new determination of the diameter of Pluto. *J. R. Astron. Soc. Canada*, 57, 163 (1963).
Halliday, I. A possible occultation by the planet Pluto. *Sky and Telescope*, 29, 216 (1965a).
Halliday, I. Pluto's diameter. *Sky and Telescope*, 30, 213 (1965b).
Halliday, I. Comments on the mean density of Pluto. *Publ. Astron. Soc. Pac.*, 81, 285 (1969).
Halliday, I.; Hardie, R.H.; Franz, O.G. & Priser, J.B. An upper limit for the diameter of Pluto. *Publ. Astron. Soc. Pac.*, 78, 113 (1966).
Hardie, R.H. Pluto's rotation and diameter. *Sky and Telescope*, 29, 141 (1965a).
Hardie, R.H. A re-examination of the light variation of Pluto. *Astron. J.*, 70, 140 (1965b).
Hardie, R.H. Pluto's dimming not permanent. *Sci. News*, 95, 285 (1969).
Harrington, R.S. & van Flandern, T.C. (In press). 1978.
Harris, D.L. In: *Planets and Satellites*, G.P. Kuiper and B.M. Middlehurst, Eds., p. 286. Univ. of Chicago Press, Chicago, 1961.
Hart, M.H. A possible atmosphere for Pluto. *Icarus*, 21, 242 (1974).
Harvard College Observatory Announcement Cards, Nos. 112, 113, 115, 117, 118, 119, 120, 121, 133.
Hatanaka, Y.; Kikuchi, S.; Konno, M. An expected occultation of a star by Pluto. I. A photoelectric observation. *Tokyo Astron. Bull.*, 226, 2623 (1973).

Herschel, J.F.W. *Outlines of Astronomy*. Longmans, Green & Co., London, 1849.
Herschel, W. Account of a comet. *Phil. Transact. R. Soc. (London)*, 71, 492 (1781).
Horedt, G. Mass loss in the plane circular restricted three-body problem: application to the origin of the Trojans and of Pluto. *Icarus*, 23, 459 (1974).
Hughes, D.W. Pluto's satellite. *Nature*, 274, 309 (1978).
Jackson, J. Neptune's orbit. *Mon. Not. R. Astron. Soc.*, 90, 728 (1930).
Jekhowsky, B. Orbit of the trans-Neptunian body. *C. R. Acad. Sci. (Paris)*, 190, 1049 (1930).
Kaplan, G.H.; Seidelmann, P.K. & Smith, E. Astrometric ephemeris of Pluto 1970-1990. *U.S. Naval Observ. Circ.*, 139 (1972).
Kelsey, L.A. & Fix, J.D. Linear polarization measurements of Pluto. *Bull. Am. Astron. Soc.*, 4, 321 (1972).
Kelsey, L.A. & Fix, J.D. Polarimetry of Pluto. *Astrophys. J.*, 184, 633 (1973).
Kelsey, L.A.; Fix, J.D. & Neff, J.S. Spectrophotometry of Pluto. *Bull. Am. Astron. Soc.*, 4, 321 (1972).
Kiang, T. The past orbit of Halley's Comet. *Mem. R. Astron. Soc.*, 76, 27 (1972).
Kiang, T. The cause of the residuals of the motion of Halley's comet. *Mon. Not. R. Astron. Soc.*, 162, 271 (1973).
Kiladze, R.I. Physical parameters of Pluto. *Sol. Sys. Res.*, 1, 173 (1967).
Kiladze, R.I. What is the mass of Pluto? *J. Brit. Astron. Assoc.*, 78, 124 (1968).
Kordylewsky, J. *Acta Astron.*, 6, 203 (1956); 7, 156 (1957); 8, 185 (1958).
Kourganoff, V. La part de la mecanique celeste dans la decouverte Plutoh.... *Bull. Astron.* 12, 147 (1941).
Kranjc, A. Approximate elements for orbits of planets of the solar system. *Mem. Soc. Astron. Ital.*, 34, 295 (1963).
Kranjc, A. Calculations of the equatorial coordinates (1950.0) of the five outer planets, and of the nutation. *Mem. Soc. Astron. Ital.*, 35, 325 (1964).
Kuiper, G.P. The diameter of Pluto. *Publ. Astron. Soc. Pac.*, 62, 133 (1950).
Kuiper, G.P. Satellites, comets, and inter-planetary material. *Proc. Nat. Acad. Sci.*, 39, 1156 (1953).
Kuiper, G.P. The formation of the planets, part III. *J. R. Astron. Soc. Canada*, 50, 170 (1956).
Kuiper, G.P. Further studies on the origin of Pluto. *Astrophys. J.*, 125, 287 (1957).

Lacis, A.A. & Fix, J.D. An analysis of the light curve of Pluto. *Astrophys. J.*, 174, 449 (1972).
Lane, W.A.; Neff, J.S. & Fix, J.D. A measurement of the relative reflectance of Pluto at 0.86 micron. *Publ. Astron. Soc. Pac.*, 88, 77 (1976).
Lau, H.E. La planete transneptunienne. *Bull. Soc. Astron. France*, 28, 276 (1914).
Lawton, A.T. Planets beyond Pluto. *Spaceflight*, 14, 454 (1972).
Lawton, A.T. "Charon" - A companion to Pluto. *Spaceflight*, 20, 428 (1978).
Lawton, A.T. The many shades of the 10th planet. *Spaceflight*, 21, 115 (1979).
Leverrier, U.J.J. Premier memoire sur la theorie d'Uranus. *C. R. Acad. Sci. (Paris)*, 21, 1050 (1845).
Leverrier, U.J.J. Recherches sur les mouvements d'Uranus. *C. R. Acad. Sci. (Paris)*, 22, 907 (1846a).
Leverrier, U.J.J. Sur la planete qui produit les anomalies observees dans le mouvement d'Uranus - Determination de sa masse, de son orbite et de sa position actuelle. *C. R. Acad. Sci. (Paris)*, 23, 428 & 657 (1846b).
Lexell, A.J. Circular elements of Uranus. *Acta Acad. Sci. Imperialis Petropolitanae, Mem.*, 4, 307 (1781).
Littlewood, J.E. *A Mathematician's Miscellany*. Methuen, London, 1957.
Lobkova, N.I. Method of constructing an analytical theory of Pluto's motion. *Sov. Astron. Lett.*, 1, 210 (1975).
Lowell, A.L. *Biography of Percival Lowell*, MacMillan, New York, 1935.
Lowell, P. Memoir on a trans-Neptunian planet. *Mem. Lowell Observ.*, I, 1 (1915).
Lyttleton, R.A. On the possible results of an encounter of Pluto with the Neptunian system. *Mon. Not. R. Astron. Soc.*, 97, 108 (1936).
MacDonald, G.J.F. Tidal friction. *Rev. Geophys.*, 2, 467 (1964).
Manning, P.G. Is Pluto an iron-rich planet? *Nature*, 230, 234 (1971).
Marsden, B.G. Comets and nongravitational forces. II. *Astron. J.*, 74, 720 (1969).
Marsden, B.G. Comets and nongravitational forces. III. *Astron. J.*, 75, 75 (1970).
Marsden, B.G. Comets and nongravitational forces. IV. *Astron. J.*, 76, 1135 (1971).
McCord, T.B. Dynamical evolution of the Neptunian system. *Astron. J.*, 71, 585 (1966).

Messell, K. Pluto. *Astron. Tidsskr.*, 7, 31 (1974).
Miller, J. *Lowell Observatory Observation Circular*, May 1, 1930.
Morgan, H.R. Definitive positions and proper motions of primary reference stars for Pluto. *Astron. Papers Am. Ephem. Naut. Almanac*, 11, 505 (1950).
Moseley, T.J. The magnitude of Pluto. *J. Brit. Astron. Assoc.*, 79, 129 (1969).
Nacozy, P.E. & Diehl, R.E. A semi-analytical theory of the secular perturbations of Pluto. *Bull. Am. Astron. Soc.*, 6, 205 (1974a).
Nacozy, P.E. & Diehl, R.E. On the long term motion of Pluto. *Celestial Mech.*, 8, 455 (1974b).
Nacozy, P.E. & Diehl, R.E. A semianalytical theory for the long-term motion of Pluto. *Astron. J.*, 83, 522 (1978).
Naef, R.A. *Orion*, 4, 484 (1955).
Neff, J.S.; Lane, W.A. & Fix, J.D. An investigation of the rotational period of the planet Pluto. *Publ. Astron. Soc. Pac.*, 86, 225 (1974).
Newburn, R.L. & Gulkis, S. A survey of the outer planets Jupiter, Saturn, Uranus, Neptune, Pluto and their satellites. *Space Sci. Rev.*, 14, 179 (1973).
Newcomb, S. An investigation of the orbit of Uranus. *Smithsonian Contr. Knowledge*, 19, No. 262 (1874)
Newcomb, S. *Astron. Papers Am. Ephem. Naut. Almanac*, VI, pt. 4 (1899)
Nicholson, S.B. & Mayall, N.U. Positions, orbit, and mass of Pluto. *Astrophys. J.*, 173, 1 (1931).
Nicolai, F. Letter. *Astron. Nachr.*, 13, 94 (1836).
O'Leary, B. Frequencies of occultations of stars by planets, satellites, and asteroids. *Science*, 175, 1108 (1972).
Opik, J. The survival of stray bodies in the solar system. *Ann. Acad. Sci. Fennicae AIII*, 61, 185 (1961).
Paraskevopoulos, J.S. Note on South African photographs of Pluto. *Pop. Astron.*, 38, 416 (1930).
Peirce, B. Investigation into the action of Neptune to Uranus. *Proc. Amer. Acad. Arts Sci.*, 1, 65 (1848a).
Peirce, B. Perturbations of Uranus by Neptune. *Proc. Amer. Phil. Soc.*, 5, 15 (1848b).
Pickering, W.H. A search for a planet beyond Neptune. *Harvard Annals*, 61, 113 (1909).
Pickering, W.H. The trans-Neptunian planet. *Harvard Annals*, 82, 49 (1919).
Pickering, W.H. The next planet beyond Neptune. *Pop. Astron.*, 36, 143 (1928a).

Pickering, W.H. The next planet beyond Neptune, Part II. *Pop. Astron.*, 36, 218 (1928b).
Pickering, W.H. The orbit of Uranus. *Pop. Astron.*, 36, 353 (1928c).
Pickering, W.H. The three outer planets beyond Neptune. *Pop. Astron.*, 36, 417 (1928d).
Pickering, W.H. Planet O. *Pop. Astron.*, 37, 135 (1929).
Pickering, W.H. The trans-Neptunian planet. *Pop. Astron.*, 38, 285 (1930a).
Pickering, W.H. The trans-Neptunian planet (supplementary note). *Pop. Astron.*, 38, 293 (1930b).
Pickering, W.H. The trans-Neptunian comet. *Pop. Astron.*, 38, 341 (1930c).
Pickering, W.H. The mass and density of Pluto. *Pop. Astron.*, 39, 2 (1931a).
Pickering, W.H. Planet P, Comet 1930 III, Wilk, Number 590. *Pop. Astron.*, 39, 321 (1931b).
Pickering, W.H. Planet P, its orbit, position and magnitude. Planets S and T. *Pop. Astron.*, 39, 385 (1931c).
Pickering, W.H. The location of Planet P (continued). *Pop. Astron.*, 39, 583 (1931d).
Pickering, W.H. The discovery of Pluto. *Mon. Not. R. Astron. Soc.*, 91, 812 (1931e).
Pickering, W.H. Planet U, and the orbits of Saturn and Jupiter. *Pop. Astron.*, 40, 69 (1932a).
Pickering, W.H. First report on the search for Planet P. *Pop. Astron.*, 40, 351 (1932b).
Pickering, W.H. A reply to Professor Brown's criticism of my views on Pluto. *Pop. Astron.*, 40, 519 (1932c).
Pickering, W.H. Pluto. A discussion of Dr. Jackson's orbit of Neptune. *Pop. Astron.*, 41, 556 (1933).
Pickering, W.H. The difference between the discoveries of Neptune and Pluto. *Publ. Astron. Soc. Pac.*, 46, (1934).
Pierucci, M. Empirical law discovered in the solar system tested by the new planet. *N. Cimento*, 7, 367 (1930a).
Pierucci, M. Orbit of the trans-Neptunian planet. *Accad. Lincei, Atti*, 11, 1091 (1930b).
Pierucci, M. Orbit of the trans-Neptunian planet. *Accad. Lincei, Atti*, 12, 103 (1930c).
Putnam, R.L. & Slipher, V.M. Searching out Pluto - Lowell's trans-Neptunian planet X. *Scientific Mon.*, 34, 5 (1932).
Rabe, E. The Trojans as escaped satellites of Jupiter. *Astron. J.*, 59, 433 (1954).
Rabe, E. On the origin of Pluto and the masses of the proto-planets. *Astrophys. J.*, 125, 290 (1957a).

Rabe, E. Further studies on the orbital development of Pluto. *Astrophys. J.*, 126, 240 (1957b).
Rawlins, D. The great unexplained residual in the orbit of Neptune. *Astron. J.*, 75, 856 (1970).
Rawlins, D. & Hammerton, M. Mass and position limits for a hypothetical tenth planet of the solar system. *Mon. Not. R. Astron. Soc.*, 162, 261 (1973).
Reaves, G. Kourganoff's contributions to the history of the discovery of Pluto. *Publ. Astron. Soc. Pac.*, 63, 49 (1951).
Reilly, T.H. Scientific objectives for imaging experiments at the outer planets. NASA Report CR-109871, Washington, 1970.
Roberts, I. *Mon. Not. R. Astron. Soc.*, (1892).
Roemer, E. & Lloyd, R.E. Observations of comets, minor planets, and satellites. *Astron. J.*, 71, 443 (1966).
Roure, H. Long period inequality in mean motion of Pluto. *C. R. Acad. Sci. (Paris)*, 198, 901 (1934).
Roure, H. Perturbations of Pluto by Neptune. *C. R. Acad. Sci. (Paris)*, 200, 437 (1935a).
Roure, H. Mean movement of Pluto. *C. R. Acad. Sci. (Paris)*, 201, 1322 (1935b).
Roure, H. Secular inequality in mean motion of Pluto. *C. R. Acad. Sci. (Paris)*, 209, 788 (1939).
Roure, H. Eccentricity of orbit of Pluto. *C. R. Acad. Sci. (Paris)*, 210, 136 (1940).
Rubashevskii, A.A. A method for the determination of the diameter of Pluto from observations of occultations of stars. *Soviet Astron.*, 10, 124 (1966).
Russell, H.N. On the light-variations of asteroids and satellites. *Astrophys. J.*, 24, 1 (1906).
Russell, H.N. Planet X. *Scientific Am.*, 143, 20 (1930a).
Russell, H.N. How Pluto's orbit was figured out. *Scientific Am.*, 143, 364 (1930b).
Russell, H.N. More about Pluto. *Scientific Am.*, 143, 446 (1930c).
Russell, H.N. Refining Pluto's orbit. *Scientific Am.*, 144, 90 (1931).
Sanders, W.L. A near occultation of a star by Pluto. *Publ. Astron. Soc. Pac.*, 77, 298 (1965).
Schuette, C.H. Two new families of comets. *Pop. Astron.*, 57, 176 (1949).
Seagrave, F.E. Circular orbit elements of the trans-Neptunian planet X. *Pop. Astron.*, 38, 355 (1930).
Seidelmann, P.K. A dynamical search for a trans-Plutonian planet. *Astron. J.*, 76, 740 (1971).
Seidelmann, P.K.; Klepczynski, W.J.; Duncombe, R.L. & Jackson, E.S. Determination of the mass of Pluto. *Astron. J.*, 76,

488 (1971).
Shapely, H. Trans-Neptunian planet. *Pop. Astron.*, 38, 189 (1930a).
Shapely, H. Trans-Neptunian planet. *Pop. Astron.*, 38, 296 (1930b).
Sharaf, S.G. Improvement of the elements of Pluto. *Trudy Inst. Theoret. Astron.*, 4, 79 (1955).
Sharaf, S.G. & Budnikova, N.A. *Trudy Inst. Theoret, Astron.*, 10 (1964).
Shechtman, D. Pluto's advance of perihelion. *Pop. Astron.*, 53, 42 (1945).
Silva, G. Il pianeta transnettuniano Plutone. *Publ. R. Osservat. Astron. Padova*, 19 (1931a).
Silva, G. Il calcoli d'orbita e l'orbita di Plutone. *Publ. R. Osservat. Astron. Padova*, 21 (1931b).
Slipher, V.M. A trans-Neptunian planet. *Pop. Astron.*, 38, 187 (1930a).
Slipher, V.M. The trans-Neptunian planet. *Pop. Astron.*, 38, 415 (1930b).
Slipher, V.M. Report of Lowell Observatory for 1930. *Pop. Astron.*, 39, 204 (1931).
Slipher, V.M. Trans-Neptunian planet search. *Am. Phil. Soc. Proc.*, 79, 435 (1938).
Smart, W.M. John Couch Adams and the discovery of Neptune. *Occas. Not. R. Astron. Soc.*, 2, 56 (1947).
Stoyko, N. Orbit of the trans-Neptunian body discovered at Lowell Observatory. *C. R. Acad. Sci. (Paris)*, 190, 1275 (1930a).
Stoyko, N. Influence of 3rd and 4th order terms in using Esclangon's method to determine a planet's orbit. Application to the trans-Neptunian planet. *C. R. Acad. Sci. (Paris)*, 190, 1379 (1930b).
Taffara, L. *Il nuovo pianeta ultranettuniano "Plutone" fotografato nel R. Osservatorio astrofisico di Catania....* M. Ponzio, Pavia, 1930.
Thomsen, B. & Ables, H.D. Measurement of the angular separation and magnitude difference for the Pluto/Charon system. *Bull. Am. Astron. Soc.*, 10, 586 (1978).
Todd, D.P. Preliminary account of a speculative and practical search for a trans-Neptunian planet. *Amer. J. Sci.*, 20, 232 (1880).
Tombaugh, C.W. In: *Planets and Satellites*, G.P. Kuiper and B.M. Middlehurst, Eds., p. 12. Univ. of Chicago Press, Chicago, 1961.
Valz, J.E.B. Letter. *C. R. Acad. Sci. (Paris)*, 1, 130 (1835).

Veronnet, A. Determination of the orbit of the trans-Neptunian planet by three observations. *C. R. Acad. Sci. (Paris)*, 191, 24 (1930).
Veverka, J. The polarization curve and the absolute diameter of Vesta. *Icarus*, 15, 11 (1971a).
Veverka, J. Photopolarimetric observations of the minor planet Flora. *Icarus*, 15, 454 (1971b).
Veverka, J. In: *Physical Studies of Minor Planets*, T. Gehrels, Ed., p. 91. NASA SP-267, Washington, D.C., 1971c.
Walker, M.F. & Hardie, R.H. A photometric determination of the rotational period of Pluto. *Publ. Astron. Soc. Pac.*, 67, 224 (1955).
Webster, W.J.; Webster, A.C. & Webster, G.T. Interferometer observations of Uranus, Neptune and Pluto at wavelengths of 11.1 and 3.7 centimeters. *Astrophys. J.*, 174, 679 (1972).
Whipple, F.L. A comet model. I. The acceleration of Comet Encke. *Astrophys. J.*, 111, 375 (1950).
Whipple, F.L. Evidence for a comet belt beyond Neptune. *Nat. Acad. Sci. Proc.*, 51, 711 (1964).
Williams, J.G. & Benson, G.S. Resonances in the Neptune-Pluto system. *Astron. J.*, 76, 167 (1971).
Wylie, I.R. A comparison of Newcomb's tables of Neptune with observation, 1795 - 1938. *Publ. U.S. Naval Obs.*, 15, pt. 1 (1942a).
Wylie, I.R. An investigation of Newcomb's theory of Uranus. *Publ. U.S. Naval Obs.*, 15, pt. 3 (1942b).
Yeomans, D.K. Nongravitational forces affecting the motions of periodic comets Giacobini-Ziner and Borrelly. *Astron. J.*, 76, 83 (1971).
Zagar, F. *Astron. Nachr.*, 240, 335 (1930).

Name Index

Adams, J.C., 6
Airy, G.B., 6, 15
Alter, D., 42, 63, 89
Andersson, L.E., 85, 95
Ash, M.E., 77

Baade, W., 48, 56
Baldet, M., 44
Baldi, P., 105
Banachiewicz, T., 41
Benner, D.C., 112, 119
Benson, G.S., 71, 104
Bessel, F.W., 4
Biesbroeck, G. van, 71
Bode, J., 2
Bouvard, A., 4
Bower, E.C., 42, 44, 45, 48, 71
Brady, J.L., 81, 102
Brouwer, D., 65, 71, 74
Brown, E.W., 46, 47, 57
Budnikova, N.A., 71
Burckhardt, J., 4
Burney, V., 39

Cacciatore, N., 6
Cameron, A.G.W., 79
Caputo, M., 105
Challis, J., 10
Christy, J.W., 121
Clemence, G.M., 65, 71, 74
Cohen, C.J., 67, 69, 71, 102, 104
Crommelin, A.C., 30, 63
Cromwell, R.H., 112
Cruikshank, D., 110, 120

d'Arrest, H.L., 14
Dauvillier, A., 53

Diehl, R.E., 120
Duncombe, R.L., 73, 74, 76, 77

Eckert, W.J., 65, 74
Encke, J.F., 14
Esclangon, E., 40, 41

Fayet, G., 51
Fink, U., 112, 119
Fix, J.D., 85, 86, 95, 98, 104, 112
Fixmillner, P., 3
Flammarion, C., 19
Flamsteed, J., 3
Forbes, G., 19, 21, 30
Foss, A., 54, 84
Franz, O.G., 92

Gaillot, M.A., 21, 22, 30
Galle, J.G., 13
George III (King of England), 2
Goldreich, P., 66
Golitsyn, G.S., 108
Gulkis, S., 100
Gunn, E.J., 79

Halliday, I., 76, 89, 92, 94
Hammerton, M., 103
Hardie, R.H., 53, 64, 85, 92
Harrington, R.S., 121, 126
Harris, D.L., 112
Hart, M.H., 106, 110
Herschel, J., 14, 22
Herschel, W., 2
Horedt, G., 105

Name Index

Hubbard, E.C., 67, 69, 71, 102, 104
Hughes, D.W., 121
Hussey, T., 6

Jackson, J., 48, 49
Jeans, J., 39
Jekhowsky, B., 40

Kelsey, L.A., 86, 98
Kiang, T., 102
Kiladze, R.I., 78
Klepczynski, W.J., 73, 74, 76, 77
Kordylewsky, J., 89
Kourganoff, V., 46, 57
Kuiper, G.P., 56, 64, 66

Lacis, A.A., 85
Lane, W.A., 104, 112
Lassell, W., 9, 15
Lau, H.E., 22, 30
Lawton, A.T., 125, 126
Le Francais de Lalande, J.J., 2
Lemonnier, P., 4
Leverrier, U.J.J., 5, 10, 11
Lexell, A.J., 2
Lowell, A.L., 30
Lowell, C.S., 35
Lowell, G., 25, 30
Lowell, P., 24
Lyttleton, R.A., 51, 66

McCord, T.B., 66
MacDonald, G.J.F., 105
Manning, P.G., 86
Marsden, B.G., 103
Maskelyne, N., 2
Mayall, N.C., 49
Mechain, P., 2
Messier, C., 2
Miller, J., 41
Morrison, D., 110, 120
Moseley, T.J., 53

Nacozy, P.E., 120
Naef, R.A., 79
Neff, J.S., 86, 104, 112
Newburn, R.L., 100
Newcomb, S., 16
Nicholai, F., 6
Nicholson, S.B., 49

Oesterwinter, C., 69, 71
O'Leary, B., 94

Paraskevopoulos, J.S., 40
Peirce, B., 16
Pickering, W.H., 22, 30, 32, 33, 39, 44, 47, 48, 49, 51
Pilcher, C.B., 110, 120
Priser, J.B., 92
Putnam, R.L., 53

Rabe, E., 64, 65
Rawlins, D., 79, 103
Reaves, G., 57
Roberts, I., 21
Roure, H., 51
Rubashevskii, A.A., 86, 94
Russell, H.N., 49

Sanders, W.L., 94
Schuette, C.H., 51
Seidelmann, P.K., 73, 74, 76, 77, 79
Shapely, H., 40
Shapiro, I.I., 77
Sharaf, S.G., 71
Shawe-Taylor, L., 54, 84
Shechtman, D., 54
Slipher, V.W., 39, 42, 53
Smith, W.B., 77
Soter, S., 66
Stoyko, N., 41
Struve, G.W., 14

Todd, D.P., 21, 30
Tombaugh, C.W., 33, 54, 79

Valz, J.E.B., 6
Van Flandern, T.C., 126

Walker, M.F., 64, 66
Webster, A.C., 86
Webster, G.T., 86
Webster, W.J., 86
Whipple, F.L., 44, 66, 79, 102
Whitworth, D., 54, 84
Williams, J.G., 71, 104

Yeomans, D.K., 103

Subject Index

Albedo, 44, 100 of Pluto, 56, 85, 108, 110, 112
Aphelia cometary, 20, 32
   of comets, 19
   of families of comets, 51
Argument of perihelion, 69, 104, 105
Atmosphere, 56, 102, 106
Axis, semi-major, 41, 69

Berlin Observatory, 13
Bode's law, 11
Brady's hypothetical planet, 84
Brera Observatory, 40
British Association for the Advancement of Science, 6

Charon, 125
Comet, 1, 33
Comet Olbers, 81
Comet Pons-Brooks, 81
Cometary aphelia, 32
Cometary belt, 79
Commensurability, 71
   with Neptune, 65
Cronos, 39

Deimos, 129
Density, 77
Diameter, 44
   of Pluto, 63, 89, 112
Dominion Observatory, 92

Eccentricity, 41, 69, 102, 105, 120
Einstein's prediction of the advance of the perihelion of Mercury, 54

Encounters of Pluto with the Neptunian system, 51

French Academy of Sciences, 10

Halley's comet, 6, 81, 102
Harvard College Observatory, 35
Heidelberg Observatory, 41
Hypothetical planets, Brady's 84
Hypothetical planets, $S1$, $P1$ and $P2$, 79
Hypothetical tenth planet, 102, 126
Hypothetical trans-Plutonian planet, 94

Iapetus, 103
Inclination, 40, 41, 69 102, 105, 120

Jupiter, 3, 11

Kitt Peak National Observatory, 98

Libration, 71, 73, 105
   of the close approach positions, 68
   of the close approaches, 69
Libratory motion, 120
Lick Observatory, 42
Limb darkening, 42
Longitude
   of the node, 40, 41
   of the perihelion, 41

Subject Index

Lowell Observatory, 30, 33
Lowell Observatory Object, 44

Magnitude, 44, 56, 85
Mass, 112
  of Pluto, 63, 74, 77, 78, 79
McDonald Observatory, 56
Mercury, 54
Methane, 108, 110, 111, 119, 120
Minerva, 39
Mount Wilson Observatory, 30

National Radio Astronomy Observatory, 86
Near-occulation, 94
Neon, 106, 110
Neon atmosphere, 108
Neptune, 14, 16
Nereid, 103

Oberon, 103
Observations, 3
Observatories
  Berlin, 13
  Brera, 40
  Dominion, 92
  Harvard College, 35
  Heidelberg, 41
  Kitt Peak National, 98
  Lick, 42
  Lowell, 30, 33
  McDonald, 56
  Mt. Wilson, 30
  National Radio Astronomy, 86
  Paris, 40
  Potsdam, 40
  Pulkovo, 14
  Steward, 116
  Uccle, 45
  Yale, 38
  Yerkes, 39
Occult, 63

Occultation, 86, 92, 94
  of a star, 89
Occultations, 91
Osculating restricted problem of three bodies, 64

Paris Observatory, 40
Percival, 39
Perihelion, 120
Perturbations of Jupiter and Saturn, 51
Perturbations of Neptune and Uranus, 49
Phobos, 129
Pioneer 10, 128
Pioneer 11, 76, 128
Plane circular restricted three-body problem, 105
Planet O, 22
Planet P, 32
Planet S, 32
Planet X, 25, 30, 38, 126
Planets
  Brady's hypothetical, 84
  hypothetical, $S1$, $P1$, $P2$, 79
  hypothetical tenth, 103, 126
  hypothetical trans-Plutonian, 94

  Jupiter, 3, 11, 51
  Neptune, 14, 16, 49
  Planet O 22
  Planet P 32
  Planet S 32
  Planet X 25, 30, 38, 126
  Pluto, 39
  Saturn, 3, 11, 51
  trans-Neptunian, 19, 22
  trans-Plutonian, 81, 102
  Uranus, 2, 3, 11, 32, 49
Pluto, 39
  albedo of, 56, 86, 108
  axis of rotation, 85
  color of, 53, 95

Pluto, 39 (cont'd.)
  density of, 42, 74, 76,
    78, 79
  diameter of, 42, 44, 74,
    77
  magnitude of, 53, 56, 85
  mass of, 42, 44, 48, 49
  radius of, 78
  rotational axis, 86, 95
Potsdam Observatory, 40
Prediscovery, 3
Prediscovery observations,
  5
Preliminary elements, 44
Pulkovo Observatory, 14

Rotation period, 64
  of Pluto, 56
Rotational period, 57
Royal Astronomical Society,
  35

Satellite, 39
  to Pluto, 121
Satellites, 34
Saturn, 3, 11
Semi-major axis, 41, 69
Solar system, 1, 79, 86
Space Telescope, 127
Steward Observatory, 116
Sun, distance from, 40
Symposium on Pluto, 127

Trans-Neptunian planet, 19,
  22
Trans-Neptunian planets, 21
Transpluto, 51
Trans-Plutonian planet, 81
  102
Triton, 48, 53, 66

Uccle Observatory, 45
Uranus, 2, 3, 11, 32
  and perturbations of
    Neptune, 49

U.S. Naval Observatory, 21,
  39

Voyager 1, 76
Voyager 2, 76

Yale Observatory, 38
Yerkes Observatory, 39